CAMBRIDGE LIBRARY COLLECTION

Books of enduring scholarly value

Astronomy

From ancient times, humans have tried to understand the workings of the world around them. The roots of modern physical science go back to the very earliest mechanical devices such as levers and rollers, the mixing of paints and dyes, and the importance of the heavenly bodies in early religious observance and navigation. The physical sciences as we know them today began to emerge as independent academic subjects during the early modern period, in the work of Newton and other 'natural philosophers', and numerous sub-disciplines developed during the centuries that followed. This part of the Cambridge Library Collection is devoted to landmark publications in this area which will be of interest to historians of science concerned with individual scientists, particular discoveries, and advances in scientific method, or with the establishment and development of scientific institutions around the world.

On the Determination of the Orbits of Comets

When this book first appeared in 1793, there had been no significant work on comets published in English since Edmond Halley's death some fifty years before. In Europe the field was dominated by French astronomers such as Pingré and Laplace, but their ornate styles were often difficult to translate. In this concise monograph, Sir Henry Englefield (*c*.1752–1822) draws both on this continental work and on his own correspondence with William Herschel to produce one of the few accessible manuals in English for the computation of cometary orbits. He includes mathematical examples as new formulae are introduced, along with detailed tables and appendices. Englefield's particular interest was in the development of scientific instruments suitable for travellers – he devised a portable telescope and lent his name to the Englefield mountain barometer – and his passion for efficiency shines through in this work, still valuable to researchers in the history of astronomy and comet science.

T0254572

Cambridge University Press has long been a pioneer in the reissuing of out-of-print titles from its own backlist, producing digital reprints of books that are still sought after by scholars and students but could not be reprinted economically using traditional technology. The Cambridge Library Collection extends this activity to a wider range of books which are still of importance to researchers and professionals, either for the source material they contain, or as landmarks in the history of their academic discipline.

Drawing from the world-renowned collections in the Cambridge University Library and other partner libraries, and guided by the advice of experts in each subject area, Cambridge University Press is using state-of-the-art scanning machines in its own Printing House to capture the content of each book selected for inclusion. The files are processed to give a consistently clear, crisp image, and the books finished to the high quality standard for which the Press is recognised around the world. The latest print-on-demand technology ensures that the books will remain available indefinitely, and that orders for single or multiple copies can quickly be supplied.

The Cambridge Library Collection brings back to life books of enduring scholarly value (including out-of-copyright works originally issued by other publishers) across a wide range of disciplines in the humanities and social sciences and in science and technology.

On the Determination of the Orbits of Comets

*According to the Methods
of Father Boscovich and Mr de la Place*

H ENRY E NGLEFIELD

CAMBRIDGE
UNIVERSITY PRESS

CAMBRIDGE UNIVERSITY PRESS

Cambridge, New York, Melbourne, Madrid, Cape Town,
Singapore, São Paolo, Delhi, Mexico City

Published in the United States of America by Cambridge University Press, New York

www.cambridge.org
Information on this title: www.cambridge.org/9781108061735

© in this compilation Cambridge University Press 2013

This edition first published 1793
This digitally printed version 2013

ISBN 978-1-108-06173-5 Paperback

This book reproduces the text of the original edition. The content and language reflect
the beliefs, practices and terminology of their time, and have not been updated.

Cambridge University Press wishes to make clear that the book, unless originally published
by Cambridge, is not being republished by, in association or collaboration with, or
with the endorsement or approval of, the original publisher or its successors in title.

ON THE

DETERMINATION

OF THE

ORBITS OF COMETS,

ACCORDING TO THE METHODS OF

FATHER BOSCOVICH AND MR. DE LA PLACE.

WITH

NEW AND COMPLETE

TABLES;

AND EXAMPLES OF THE CALCULATION BY BOTH METHODS.

BY SIR HENRY ENGLEFIELD, BART.
F.R.S. & F.A.S.

NOS QUOQUE SUB DUCIBUS CŒLUM METABIMUR ILLIS.

LONDON:

PRINTED BY RITCHIE AND SAMMELLS,

FOR PETER ELMSLY, IN THE STRAND.

MDCCXCIII.

TO

THE RIGHT HONOURABLE

JAMES STUART MACKENZIE:

AS A MARK OF SINCERE ATTACHMENT,

AND RESPECTFUL GRATITUDE

FOR FRIENDSHIP

CONTINUED FROM HIS FATHER TO HIMSELF,

THIS WORK

IS INSCRIBED,

BY HIS FAITHFUL AND AFFECTIONATE

HENRY C. ENGLEFIELD.

PREFACE.

THE Theory of Comets, difcovered and demonftrated by Newton, has not only been the bafis of all fubfequent refearches in that curious and difficult branch of Aftronomy; but in its great principles and effential parts, was brought to abfolute perfection by its illuftrious Author. Having proved that Comets, paffing through the regions of fpace in all directions, and traverfing the whole folar fyftem, do many of them approach much nearer to the Sun than any of the Planets; he attempted the inveftigation of the real path of a Comet, from its places obferved on the Earth. This Problem, which even he calls, *Longè difficillimum*, he completely folved; and demonftrated that Comets move round the Sun in orbits extremely excentric, but obeying the fame laws which guide the Planets of the Solar Syftem.

Dr. Halley, whofe affiduous labours in every branch of Aftronomy during a life protracted to a more than ufual extent, will ever be fubjects of admiration; notwithftanding the immenfe length and labour of the Newtonian procefs, applied it to twenty-four Comets, every one of which afforded an additional and complete confirmation of the truth of Newton's Theory, and the accuracy of his method.

But though the precifion of this method left nothing to be defired on that head; yet its extreme length and difficulty induced fubfequent Mathematicians to feek for modes of fhortening and fimplifying this laborious procefs. Almoft every Academy has propofed fome branch of the Theory of Comets as a fubject for their Prize differtations; and the genius of almoft every eminent Geometer on the Continent, has at different times been exercifed on this curious and interefting fubject.

This country has however unaccountably neglected the Aftronomy of Comets, and fince the deceafe of Dr. Halley, nothing has been publifhed on this fubject worthy of mention, except the excellent little treatife of Mr. Barker, printed in the year 1757:* while in Italy, Father Bofcovich; in Germany, Lambert and Tempelhoff; in Holland, Struyjk; in France, among many others, Clairaut, Du Sejour, De la Grange, and De la Place; in Ruffia, the great Euler, and Lexell; have at differ-

* This work, entitled " An Account of the Difcoveries concerning Comets," contains not only a very accurate and practical account of the Newtonian Method of Computation; but a Table of the Parabola, which in extent, accuracy and convenience, is equal to any extant; and till the publication of Mr. Pingrè's great table of the Parabola in his Comêtographie, was far fuperior to any in ufe: yet this excellent book was fo totally unknown on the Continent, that neither Mr. Mechain nor Mr. Pingrè had ever heard of it till it was made known to them by me. Mr. Mechain in a letter to me fpeaks in the higheft terms of the Table of the Parabola.

ent times contributed largely to clear the roads or extend the limits of knowledge on this fubject.

Their writings however, forming part of voluminous Academic Collections, were acceffible to few; or printed in fmall tracts were in danger of being entirely loft; till Mr. Pingrè collected thefe fcattered Rays into one Focus in his great Work, called *Cometographie*; in which his abilities as an Hiftorian, and Geometer, his deep Refearch, and critical Skill, are equally apparent. Yet this excellent Work from its fize neceffarily expenfive, and written in a foreign language, cannot be fo generally known in this country as it deferves; and we are yet without a book in our own tongue, which in a moderate compafs may make us acquainted with the modern methods of Cometary Calculation, moft generally ufed and approved, by thofe who have peculiarly applied themfelves to this branch of Aftronomy.

This deficiency the following Pages are intended in fome meafure to fupply; not by giving to the Public a complete Tranflation of the Mathematical part of Mr. Pingrè's work, a tafk far above my abilities, and which would certainly be too expenfive to anfwer my purpofe; but by a full detail of the two methods which I have tried with fuccefs; that of Father Bofcovich for a firft approximation to the Elements, which will almoft always enable the Computer, after a few days obfervations of a Comet, to predict its motion fo as to find it after an interval of bad weather, or Moonlight, or proximity to the Sun*;

* As a proof of the degree of Accuracy generally attainable by the Method of Father Bofcovich, the following Comparifon is given of the Approximate Elements of two late Comets, as determined by me after

and that of Mr. De la Place, both for a firſt Approximation, and the final determination of Elements as exact as the Obſervations themſelves will allow.

Although however I have ſaid that both theſe Methods are given in the Cometographie of Mr. Pingrè, yet it is not from that Book, but from the originals of the Authors that the preſent work has been compiled. The Method of Father Boſcovich which he firſt publiſhed in two moſt elegant Latin Diſſer-

a very few days Obſervations; with the correct Elements computed by Mr. Mechain, from the whole of the Obſervations on thoſe Comets.

	ENGLEFIELD.				MECHAIN.		
	1790.						
	s	o	'	"	s	'	"
Longitude of Aſcending Node	1	5	14	0	1 . 3	11	2
Inclination of the Orbit		63	35	0		63 . 52	27
Place of Perihelion on Orbit	9	4	57	20	9 . 3	43	27
Logarithm of Perihelion Diſt.	9.8981795				9.9019814		
	0.791005				0.797960		
Time of arrival at Perihelion	D	H	M	S	D	H	M	S
May	20	11	21	0	21	5	47	0
Motion Retrograde.							

	1792.						
	s	o	'	"	s	o	'	"
Longitude of Aſcending Node	6	11	55	0	6	10	46	15
Inclination of the Orbit		41	5	0		39	46	55
Place of Perihelion on Orbit	1	4	43	0	1	6	29	42
Logarithm of Perihelion Diſt.	0.1111953				0.1116064		
	1.2918				1.29301		
Time of arrival at Perihelion	D	H	M	S	D	H	M	S
January	15	6	0	0	13	13	35	0
Motion Retrograde.							

tations inferted in the 7th Volume of the *Memoires des Scàvans Etrangers*, is very concifely given, from thofe papers, in Mr. Pingrè's work; fo concifely indeed, as hardly to do juftice to the method as there given by the Author; but fince the impreffion of that part of the Cometographie, the learned Author made feveral improvements in his method, and publifhed the whole in a very long treatife, which makes a part of a work in five Quarto Volumes, intitled *Opufcula*; printed at Baffano in the year 1785, but a fhort time before his death: this treatife is written in very indifferent French, full of repetition and fuperfluous detail; in which the effential parts of the Method are fo buried, as not to be followed without much pains and difficulty. It therefore became neceffary not to tranflate, but to extract and recompofe many parts of it; and I am yet afraid that I have not been able wholly to correct the defects of the Original. I hope however that I have not added any miftakes of my own. Some obfervations which occurred in the courfe of the many trials made of the Method, I have ventured to give in the form of notes; and in order to render a reference to the Original, eafy to any perfon who may be inclined to compare this work with it; to each paragraph is added the number of that fection to which it correfponds in the French text.

The learned and valuable Method of Mr. De la Place is given by him in a Memoir in the Volume of the *Memoires de l'Academie des Sciences*, for the Year 1780; and confifts of two parts; the firft contains the Principles on which his method is founded; the fecond, the practical part of it.

This work only contains the Second part of Mr. De la Place's Memoir, in tranflating which, I have not exactly followed his

Text; but have combined it with Mr. Pingrè's account of it, and inserted many very useful Observations and Explanations, which I owe to the friendship of Mr. Mechain. An example of the whole process, in both the Methods of Cometary Calculation, has been added; with every part of the Computation at full length; such examples being of much use in giving a distinct idea of the rules. To the rule however for finding the Comet's Geocentric place from the Elements given, no example is given; the process being very simple, and to be found in every book which treats on the subject.

The Tables which are added, are all which can be wanted in this part of Cometary Calculation: and are the most extensive and accurate extant. Full instructions for their use, with complete examples, are given.

It was at first intended to have given an history of the principal Comets observed since the discovery of Telescopes, particularly with regard to the Phœnomena of their Nuclei and Tails; but the work has by degrees increased to such a size, that any further addition to it would have rendered it more expensive than was wished, as the principal aim was to make it of general use. Should however the public at all encourage the further prosecution of the subject, much curious and singular detail is collected both from scarce books and Manuscripts, which may at some future time be communicated, as a second part of this work.

Some alterations and improvements which occurred since the former sheets of the work were printed off, are given at the end of the book as an appendix. The Table of Abscissæ and Ordinates, which may sometimes be very convenient for drawing

Parabolas, and which, as far as I know, has never been given; is added to the other Tables. Its uſes are explained in the Appendix.

Every endeavour has been uſed to render the impreſſion of this work, as correct as poſſible: ſome errors it is however ſcarcely poſſible to avoid. The Table of Errata is the reſult of repeated examinations of the book; and it is hoped that no error is left undetected.

CONTENTS.

CHAPTER I.

General View of the Method.

§ $\frac{1}{1}$. **I**N figure 1,* let S. be the Sun; T, T′, T″, the places Fig. 1 of the Earth at the three obfervations; C, C′, C″, the correfponding places of the Comet; P, P′, P″, the fame places projected on the plane of the Ecliptic; t, c, p, the interfections of the Radii S T′, S C′, S P′, with the cords T T″, C C″, P P″. C I, is a right line parallel and equal to P P″, which meets P″ C″, (prolonged if neceffary) in I. I fuppofe the obfervations fo near one another, that the verfed fines T′ t, C′ c, P′ p, are fmall, compared to their Radii; yet far enough for the motion of the Comet in longitude not to be infenfible, or fo fmall as to be materially affected by fmall errors in obfervation. In general, an interval of from five to ten days between each obfervation, will be the beft; but in fome cafes, an interval of a day and an half, will not be too fmall, and fometimes, an interval of fifteen days, will not be too great.

§ $\frac{2}{2}$. Obfervation gives the three longitudes of the Comet determined by the directions T P, T′ P′, T″ P″; and three

---

* In this figure, the triangular fpace, P″ C″ R, muft be conceived, as ftanding at right angles to the plane P″ R S; and of courfe the angles, C P T, C P S, C′ P′ T′, C′ P′ S, and the reft; to be right angles.

Fig. 1. latitudes P T C, P′ T′ C′, P″ T″ C″. The Ephemeris gives the three longitudes of the Earth, determined by the lines S T, S T′, S T″, with their diftances from the Sun. The theory of Aftronomy gives the excentricity of the Earth's orbit o , 017 of the mean diftance, and by the Newtonian theory of univerfal gravitation, we know that all bodies moving round the Sun defcribe areas proportional to the times ; and that the fquare of the velocity in the Parabola, is reciprocally proportional to the diftance of the Sun ; being double the fquare of the velocity in the circle, at equal diftances from the Sun.

§ $\overset{2}{3}$. From thefe data, and the geometrical properties of the Parabola, we fhall determine the dimenfions and pofition of the Parabola, defcribed by the Comet, and the point in whicn it is in its orbit at a given time ; its diftance from the Sun, and the Earth ; and its heliocentric longitude, and latitude.

§ $\overset{3}{4}$. The principal foundation of this method is the fub-ftitution of an equable and rectilinear motion of the Comet in the cord of its orbit, to its curvilinear and unequal one in its orbit.

§ $\overset{4}{5}$. We firft find that when the areas are proportional to the times, though the velocity of motion in the curve is very unequal ; the velocity of the interfection of the Radius Vector and the Cord is very nearly equable, in arches whofe verfed fine is fmall, compared with the Radius Vector. Hence, by fubftituting the interfections of the Radius Vector of the Comet, and the Earth, at the fecond obfervation, to the points of their orbits ; we deduce the ratio of the curtate diftances of the Comet from the Earth, to one another. One of thefe dif-tances being affumed as known, gives the others ; from thefe we find the Radii Vectores, (or true diftances of the Comet from the Sun,) and the Cord of the Parabolic Orbit defcribed by the Comet. Another Theorem gives the Ratio of the fum of the Radii Vectores and the Cord to the motion of the Earth in its orbit, during the time elapfed between the firft and laft obfervation. The application of this Theorem to the Radii and Cord, deduced from the firft affumed curtate diftance from the Earth, gives the error of that pofition.

§ 6.⁴ The fubftitution of thefe interfections changes the longitude obferved in the line T′ P′, into a longitude which would be determined by the line t p. But firft it is eafy to fee that when the Comet is in conjunction with or in oppofition to the Sun, thefe two lines fall into one; and I demonftrate that when the Comet and Earth are equidiftant from the Sun, they are parallel: in thefe circumftances, which are not rare, no reduction for the fubftitution of one longitude for another, is neceffary: and in pofitions not much different from them, the reduction is fo fmall as to be fafely neglected. A very near approximation may indeed be often obtained, in pofitions far diftant from thefe, without employing the reduction.

§ 7.⁵ A fhort and fimple procefs gives this reduction, which enables us, without fenfible error, to fubftitute the motion of the interfection of the Radius Vector with the Cord, to the motion in the Arch. To find this reduction, we employ the verfed fine of the Parabolic Arch, taken on the fecond Radius Vector. This verfed fine is found from the correfponding verfed fine of the Arch of the Earth. This laft is eafily known, and is to the firft, very nearly in the reciprocal ratio of the fquares of the diftances from the Sun; thefe two verfed fines being nearly the meafures of the effects of gravity on the two bodies.

§ 8.⁶ Having, when neceffary, made the reduction of the fecond longitude as above, the two following Theorems give

Fɪɢ. ɪ.   the Radii Vectores, and Parabolic Cord; and of courfe the figure and pofition of the Comet's orbit.

T H E O R E M   Fɪʀsᴛ.

The ratio of one of the curtate diftances of the Comet from the Earth, to any other curtate diftance; is compofed of the direct ratio of the intervals of time between the two obferva-tions belonging to thofe diftances, and the third obfervation; and of the reciprocal ratio of the motions in longitude anfwer-ing to that time.

T H E O R E M   Sᴇᴄᴏɴᴅ.

§. $\frac{7}{9}$ If a be the fquare of twice the fpace which the Earth would pafs through by its mean motion in the interval of time between the firft and third obfervation;

b, the fum of the two extreme diftances of the Comet from the fun;

c, the Cord of the Arch of the Comet's motion between the firft and third obfervation: then $b c^2 = a$, when the Parabolic Arch is very fmall.

And when it is rather greater, a fmall correction is neceffary to the firft term, and the equation will ftand $b c^2 - \frac{c^4}{12b} = a$.

This correction is deduced from the fmall inequality found in the velocity of the interfection, when it is compared with the velocity of the Comet in the Parabolic Arch. The little inequality of the former is determined, and from thence the mean velocity of the interfection, and the point of the Arch in FIG. 1. which the Comet has a velocity equal to this mean velocity, are found. The value of a, is eafily found by a conftant logarithm, and the double logarithm of the total time reduced into minutes.

§ $\overline{10}$. Let us now draw a circle of any fize, for the ecliptic, having the Sun in its centre, and having drawn three Radii for the longitudes of the Earth at the three obfervations, and fet off on thefe Radii the true diftances of the Earth from the Sun at each obfervation, from thefe points we draw three indefinite lines for the three obferved longitudes of the Comet; each of . them making, with the Radius Vector of the Earth, an angle equal to the elongation of the Comet from the Sun, or the difference of their longitudes.

§ $\overline{11}$. For the firft pofition we affume a curtate diftance of the Comet from the Earth at the fecond obfervation, and not to err very much in this affumption, we may form a guefs from the length and pofition of the Cord C C″, of the Para- FIG. 1 & 4. bolic Arch. If the diftance of the Comet from the Sun is twice that of the Earth, this Cord will be nearly equal to the Cord T T″, of the Arch of the Earth's Orbit. If the diftance of the Comet be half that of the Earth, the Parabolic Cord will be double that of the Earth in confequence of the theory of gravitation, (§ 2.) And as the length of the Parabolic Cord varies much flower than the diftances from the Sun, a near judgment of the diftances may be formed from the eftimated Cord. The Cord P P″, and diftance S P′, reduced to the plane of the Ecliptic, are always lefs than the true Cord C C″, and Radius Vector, S C′, of the real orbit, owing to its inclination to the plane of the Ecliptic: and this difference is eafily judged of from the latitudes, which raife the points of the orbit perpendicularly over the points of projection P, P″, to C, C″.

C

Fig. 4. The position of the reduced Cord must be such, that its ends must coincide with the lines of the two extreme observed longitudes of the Comet, and be cut in the ratio of the times elapsed between the observations, in a point which must be very near the line of the second longitude; so as to leave on the Radius Vector, which cuts it in this ratio, a very small space for the versed sine.

§ 12. Set off therefore on a strait edge of a piece of paper three points of the Cord of the Earth's Arch T, t, T″, Fig. 4.* and move it on the lines of the observed longitudes of the Comet, backwards and forwards, keeping the point p near the line of the second longitude, and between it and the Sun, to leave on the opposite side, the small space answering to the versed sine of the Parabolic Arch; and so that the distances of the two points G, G′, from the two lines of extreme longitudes, shall be very nearly in the ratio of the intervals p G, p G′, going beyond the lines of longitudes with respect to p, or keeping within them, according as by the circumstances mentioned in the last paragraph, we may judge the reduced cord less or greater than the Earth's. A little habit will direct to a near guess at this position. When the daily motion of the Comet in longitude is not very small, by moving the edge a very little, the part intercepted between the lines of extreme longitudes will change much, and the limits of probability of the real length of the cord, will lie in a narrow compass. By this first supposition, we shall in general not be very far from the truth.

§ 13. But at all events, having assumed a first position, and having the point S, we have also the curtate distance from the

---

* The points T t T″ so transferred and applied to the lines of elongation of the Comet, are represented in Fig. 4. by G p G′, in the situation chosen for them after the considerations mentioned in the text.

Sun ; and by means of the angles of latitudes, prepared apart, as hereafter defcribed, the elevation of the Comet from the plane of the Ecliptic for this pofition, and its whole diftance from the Sun are eafily found ; and are the elements for the reduction of the fecond longitude. This reduction being made, by a fhort calculation, we fhall deduce from the fecond curtate diftance from the Earth, the two extreme ones, which will give the two curtate diftances from the Sun ; and by the angles of latitudes, the elevations from the plane of the Ecliptic will be found. From hence will be found the firft, and third, whole diftances from the Sun ; and the length of the cord, which give the values of b, c, in the formula $b\,c^2 - \frac{c^4}{12\,b} = a$. If the value of the formula comes out different from a, known before, the fecond curtate diftance muft be changed, and in general diminifhed, if the value is greater; and increafed, if fmaller, than a. The errors of the formula in two pofitions, will by a fimple proportion give the true pofition, or near it; and very often the fecond pofition will be fo near the truth, as to give the orbit of the Comet near enough to find its apparent motion for the reft of its appearance.

§ 14. The elements of the orbit will be eafily found, as we fhall hereafter fhew.

§ 15. If more precifion is defired, after two or three conftructed pofitions, trigonometry may be called in. And the folution of three oblique angled, and three right angled, plane triangles, will give the diftances from the Sun, and the Cord, which anfwer exactly to the pofition; and having obtained $b\,c^2 - \frac{c^4}{12\,b} = a$; a fimple trigonometrical calculation, will give all the elements of the orbit.

CHAPTER II.

On the Motion of the Point of Interſection of the Radius Vector and Cord.

FIG. I. $\overset{16}{\S \text{ I.}}$ **I**N the orbit of the Comet C C′ C″, Figure 1. let cords be drawn joining C C′, and C′ C″; the ſegments intercepted between theſe cords, and the arches of the orbit, will be very ſmall, when compared with the areas of the ſectors C S C′, C′ S C″; the ratios of theſe ſectors will therefore be ſenſibly the ſame as thoſe of the triangles C S C′, C′ S C″, which only differ from them by the ſegments; now the areas of the ſectors are as the times; and the areas of the triangles as C c, to c G″; which are the baſes of the triangles C S c, c S′ C″, having a common termination at S; and of the triangles C C′ c, c C′ C″, having a common termination at C′; therefore the ſegments C c, c C″, of the cord C C″, are ſenſibly as the times in which they are moved through by the interſection of the Radius Vector S C′. Therefore the motion of the point of interſection is ſenſibly uniform.

$\overset{17}{\S \text{ 2.}}$ The diviſion of the cord by the point of interſection in the ratios of the times, will be almoſt perfectly exact near the middle, where the triangles will be equal, and the ſegments nearly ſo. There is a point in which the ratio is exact, but without ſeeking this point, it is obvious, that in arches whoſe verſed ſine is ſmall when compared with the Radius Vector, as we ſuppoſe in theſe reſearches; when the intervals of time are nearly equal, the two ſegments which we neglect, ever very fmall, and in this cafe nearly equal, will leave an infenfible error in this ratio.

§ 3.[18] As in all orbits defcribed by central forces, the areas are in the direct ratio of the times; it follows that the theorem above ftated, holds good in them all. Therefore in the Cord T T″, of the Earth's orbit; the Segments T t, t T″, will be fenfibly in the ratio of the times. Fig. 1. & 4.

§ 4.[19] But it will be ufeful to determine the little inequality, which exifts in the velocity of the point of interfection of the Radius Vector, and the Cord; as by this means greater arches may be made ufe of. This correction is the $\frac{c^4}{12b}$ in the formula given in the laft chapter. Let then C A C′ (figure 2) be the arch of a curve, the areas of whofe fectors terminated in S, are in the ratio of the times; conceive two Radii Vectores S A, S a, infinitely near each other, and meeting the Cord C C′, in B, b; and the line S G, perpendicular to the Cord; then the triangle B S b, and the Sector A S a, may be confidered as fimilar, or as two fectors of circles; therefore the firft will be to the fecond as S B², to S A²; and as the firft is $= $ B b $\times \frac{1}{2}$ S G, the fecond will be $= \frac{\frac{1}{2} S G \times S A^2 \times B b}{S B^2}$. Now the velocity of the point of interfection B, is as the fpace B b, divided by the time employed to pafs over it; which is proportional to the Sector A S a. The ratio therefore of the velocity fought, will be found by dividing B b, by the value of the Sector, which gives $\frac{2 S B^2}{S G \times S A^2}$, and as 2, and S G, are conftant; the velocity will vary as the fraction $\frac{S B^2}{S A^2}$. Fig. 2.

D

FIG. 2.

§ 4. If the arch is defcribed by forces tending to the centre S, it will be ever concave towards that point, and if the arch is not large, it will have a tangent K D K', parallel to the Cord C C', and beyond it, fo that the Radius Vector terminating at the point of contact D, fhall cut the Cord C C', in a point F, and will cut other parallel lines A A' in points I.

The velocity in queftion, will be greateft at the points C, and C'; and will go on continually diminifhing from C to F, where it will be at its minimum; and then increafe again from F to C'; but this inequality, will be ever fmall when the verfed fine D F, is fmall, if compared to the Radius S D: for we ever have $\frac{S F}{S I^2} = \frac{S B^2}{S A^2}$; and as S F, is conftant, the velocity of the inter-fection B, will be in the reciprocal ratio of the fquare of S I; which will be the leaft when the points A, A', coincide with C, C', and the point I, falls on F; and will be greater, in pro-portion as A, A', approach to D; the line S I, having a maxi-mum, when A, A', fall on D; when it is equal to S D. as the change of the line S I, never exceeds the verfed fine D F, its variation, and the variation of the velocity, which is recipro-cally proportional to it, will be ever fmall.

§ 5. There will be a mean velocity, with which in the fame time, the fame Cord would be defcribed with an uniform velocity: we fhall now find the point Q, of the parabolic orbit, in which the Comet has a velocity, equal to that mean velocity. The length of the Radius S Q, is the principal object of this re-fearch; as from that, will be deduced the correction of the for-mula, given in Chapter 1, § 9.

This length will be firft found as compared with the Radius S D, and the verfed fine D F; from whence we fhall find its value, with refpect to the Cord C C′=*c*, and the fum of the Radii S C, S C′, which is = *b*.

§ 6. Let the lines C H, D T, C′ H′, be drawn perpendicular to the directrix of the Parabola, having its focus in S; K, K′, being the points of meeting of the two extremes, with the tangent drawn through D. From the properties of this curve, we have the following Theorems.

THEOREM FIRST.

C H, D T, C′ H′, will be equal to the Radii S C, S D, S C′, and parallel to the Axis; from whence it follows, that a perpendicular let fall from the focus, on the directrix, paffes through the vertex of the Parabola; and is twice the diftance of the vertex, from the focus.

THEOREM SECOND.

D T being prolonged forms a diameter, which has for its ordinate, the Cord C C′; and bifects C C′, in E.

THEOREM THIRD.

The two lines D E, D F, being equally inclined to the tangent K D K′, will be alfo equally inclined to the Cord C C′; and therefore will be equal.

THEOREM FOURTH.

The parameter of this diameter, will be = 4 S D; from whence the value of the two lines is deduced = $\dfrac{E C^2}{4 S D} = \dfrac{c^2}{16 S D}$.

THEOREM FIFTH.

The parameter of the axis, will be four times the diftance from the focus, to the vertex; and the femiordinate to the axis, drawn through the focus, will be double that diftance.

FIG. 2.

THEOREM SIXTH.

The perpendicular E T, will be equal to the half fum of C H, C′ H′; that is, of the two Radii S C, S C′ = $\frac{1}{2}$ b; and from thence S D, = D T, = E T = E D, = $\frac{1}{2}$ b — $\frac{c^2}{16\,S\,D}$; or by fubftituting $\frac{1}{2}$ b, for S D, in the fecond term which is fmall, S D = $\frac{1}{2}$ b — $\frac{c^2}{8\,b}$.

THEOREM SEVENTH.

The area of the Parabolic Segment C D C′, is equal to $\frac{2}{3}$ of the Parallelogram C K K′ C′; therefore if E N, be taken = $\frac{2}{3}$ E D; and through N, be drawn a line parallel to the Cord C C′, and meeting the Radius S D, in O, and the perpendiculars C H, C H′, in M and M′; the Parallelogram C M M′ C′ will be equal to the Parabolic Segment. If the fame line M M′ meets the Radii S C, S C′, in L, and L′; the fame Parallelogram may be taken, as equal to the quadrilateral fpace C L L′ C′; becaufe of the fmallnefs of the two triangles C L M, C′ L′ M′; therefore the triangle L S L′, may be taken as equal to the Parabolic Sector.

§ 7. From thefe data it follows, that if the line M O M′ meets the Radii S A, S a, in R, r; the triangle R S r, will exprefs the time of the motion through B b, with a mean velocity, while the Sector A S a, expreffes it with the actual unequal velocity. For conceive the Cord C C′ divided into an infinite number of equal parts, as B b; it is evident, that the areas of all the triangles R S r, will be equal amongft themfelves; as are thofe of the triangles B S b = $\frac{1}{2}$ B b × S G; becaufe of the conftant ratio between them, which is that of $S\,B^2$, to $S\,R^2$: or $S\,C^2$ to $S\,L^2$: having therefore expreffed the time employed in the conftant portion B b, by the triangle R S r, we fhall have that time conftant, and from thence a conftant velocity; on the other hand the total time expreffed in the motion, by the triangle L S L′; will be equal to the total time expreffed in the actual unequal motion, by the Sector C S C′; from whence it follows that the firft motion, will be the mean motion; and its velocity, the mean velocity.

§. 8. The velocities of the mean motion, and the actual mo- [25] tion at B, will be in the reciprocal ratio of the time employed in the fpace B b; therefore the mean velocity will be to the actual velocity as the fmall Sector A S a, to the triangle R S r, that is as $S A^2$ to $S R^2$. When the point A, coincides with D, the ra- tio of the mean velocity, to the actual velocity in F, will be as $S D^2$, to $S O^2$. But then, the actual velocity of the interfection B, in the Cord, will be to the actual velocity of the Comet in D, as S F, to S D; becaufe the actual velocity of the point of in- terfection, is ever to the actual velocity of the Comet, as the fmall line B b, to the fmall Arch A a; which are the fpaces cor- refponding to the fame time; and in that cafe the Cord of the fmall Arch A a, taking the direction of the tangent, will become parallel to the Cord C C′; that is; to B b; and we fhall have B b : A a :: S B : S A that is, :: S F : S D. Therefore the mean velocity of the point of interfection, will be to the actual velo- city of the Comet at D; which is compounded of thefe two ra- tios, as $S D^2 \times S F : S O^2 \times S D :: S D \times S F : S O^2$.

§ 9. Hence at length is deduced the determination of the Ra- [26] dius S Q, terminated at the point Q, in which the actual velo- city of the Comet, is equal to the mean velocity of the point of interfection of the Radius Vector, and the Cord, (§ 5). As the

E

FIG. 2. square of the velocity of the Comet is in a reciprocal ratio to its distance from the Sun, we have $SD^2 + SF^2 : SO^4 :: SD :$ $SQ = \dfrac{SO^4}{SD \times SF^2}$. As DO is $\frac{1}{3}$ of the verfed fine DF; ($\S 6$) from it, is obtained the value of the Radius SQ, fought; ($\S 5$) by the Radius SD, and its verfed fine DF.

$\S$ 10. The fmallnefs of the verfed fine, when compared with the Radius, gives a much more fimple expreffion to the Radius fought. For taking SP, as a third proportional to SD, SO, and $SV = SQ$; the ratio of the mean velocity of the interfection, to the actual velocity of the Comet at D, being compofed of thefe two $SD^2 : SO^2$, and $SF : SD$: the firft being made $= SD : SP$; the compound will become $SF : SP$. We then may fay $SD : SV = SQ :: SF^2 : SP^2$; now we have the proportion $SD : SO :: DO : OP$: and as SD, is fenfibly equal to SO, we fhall have $OP = OD = \frac{1}{3}DF$; and from thence $= FP$. Therefore SP will be fenfibly equal to the mean either Arithmetical or Geometrical between SF, SO; whence is deduced $SF : SO :: SF^2 : SP^2$; and therefore $SD : SV ::$ $SF : SO$; and $SD : DV :: SF : FO$; and as SD, is fenfibly equal to SF, we have alfo $DV = FO = \frac{2}{3}DF$. Hence is deduced the following more fimple and elegant determination of the Radius fought; deduced from the Radius SD, and its verfed fine, DF.

Add to the Radius two-thirds of its verfed fine; and the fum will be the Radius fought; that is, the diftance from the Sun, at which the Comet will have a velocity in its Orbit, equal to the mean velocity, of the point of interfection of the Radius Vector, and Cord.

§ 11.[28] A fhort calculation, would have obtained the fame refult, from the formula $SQ = \dfrac{SO^4}{SD \times SF^2}$; neglecting the terms which have the higher powers of the value x; which is $= \frac{1}{3} DF$. If SD, be $= n$; we have $DF = 3 x$, and $\dfrac{SO^4}{SD \times SF^2} =$

$$\dfrac{(n-x)^4}{n(n-3x)^2} = \dfrac{n^4 - 4n^3 x}{n^3 - 6n^2 x} = n + 2x.$$

§ 12.[29] This expreffion, gives alfo the value of the fame diftance, by the fum of the extreme Radii Vectores, SC, SC', $= b$; and by the Cord $CC' = c$, which we have propofed to deduce from them. (§ 5.) For taking, (§ 6) $SD = \frac{1}{2} b - \dfrac{c^2}{8b}$, and $DF = \dfrac{c^2}{16 SD} = \dfrac{c^2}{8b}$; we have $SQ = SD + \frac{2}{3} DF = \frac{1}{2} b -$

$$\dfrac{c^2}{8b} + \dfrac{c^2}{12b} = \frac{1}{2} b - \dfrac{3c^2}{24b} + \dfrac{2c^2}{24b} = \frac{1}{2} b - \dfrac{c^2}{24b}.$$

The following Chapter will fhew how to deduce from this equation, the formula $bc^2 - \dfrac{c^4}{12b} = a$; and how to find the value of a.

CHAPTER III

On the Comparifon of the Parabolic Cord, with the Space which anfwers to the mean Velocity of the Earth in the fame Time.

FIG. 2.   § 1.   LET u, be the fpace which the earth would defcribe in a given time, with her mean velocity, in a circle, whofe Radius is the mean diftance from the Sun $= 1$. Then the fquare of the fpace, which the Comet would pafs through in that time, at the fame diftance from the Sun; will be $= 2 u^2$, (Chapter 1 § 2). and we fhall have the proportion, $S Q = \frac{1}{2} b - \frac{c^2}{24 b} : 1 : : 2 u^2 : c^2$; becaufe that is the fpace which anfwers to the velocity of the Comet in Q, that is, to the mean velocity of the point of interfection of the Radius Vector, and the Cord C C'. We therefore have $\frac{1}{2} b c^2 - \frac{c^4}{24 b} = 2 u^2$, and making $a = 4 u^2$, we have $b c^2 - \frac{c^4}{12 b} = a$; which is the formula of Chapter 1 § 9. In this formula, the value a, is the fquare of $2 u$; that is, of double the fpace which the earth would move through with its mean velocity, in the given time. We muft now find this, from the time given.

§ 2. Let t, be the time elapfed between the firft and fecond Obfervation, reduced to minutes and tenths, t', the time between the fecond and third, and t'', the fum of t, and t'; $=$ the time elapfed between the firſt and third obſervation. Let p be the time of the Earth's annual revolution with reſpect to the fixed ſtars, reduced to ſeconds of time; let the Ratio of the Diameter of the circle to its circumference be $1 : q$; then in the Earth's Orbit, whoſe Radius is taken $= 1$, the circumference will be $2q$: then ſay $p : 2q :: 1' = 60'' : \frac{120''q}{p}$; which will be the ſpace moved by the Earth in a minute, with its mean velocity; and for t'', the total number of minutes, the ſpace will be $\frac{120''qt''}{p}$. Then the value of a, ſought, will be the ſquare of $\frac{240''qt''}{p}$

§ 3. The annual revolution of the Earth, with reſpect to the fixed Stars, is 365. 6. 9. 10 $= 31558150 = p$; the well-known ratio of the circumference of a circle to its diameter is

D. H. M. S.   S.

$3,1415927 = q$; therefore we have $\frac{240''q\,t''}{p} = \frac{24 \times 3.\,1415927\,t''}{31558150}$.

And taking the Logarithms of theſe numbers, and adding the arithmetical complement of the logarithm of the diviſor, the Logarithm of the above fraction ſtands thus :

24 Logarithm - -	1. 3802112
3. 1415927 Logarithm -	0. 4971499
31558150 Logarithm ᴀ : ᴄ :	3. 5008884
Logarithm t''	

$$\frac{24 \times 3.\,1415927\ t''}{3155815} \qquad = \qquad 5.\,3782495 + \text{Log. } t''.$$

F

FIG. 2. And the double of this Logarithm, 0. 7564990 $+$ 2 t'', is the Logarithm of a. This then is the rule for finding a. To the Conftant Logarithm 0. 7564990, add twice the Logarithm of the time elapfed between the firft and third obfervation; the fum is the Logarithm of a.

CHAPTER IV.

Of the Reduction of the Second Longitude of the Comet,
as obferved in the Arches of the Orbits of the Earth
and Comet, to that which would have been obferved in
their Cords.

FIG. I.

§ 1. $\overset{33}{}$ IT is in this chapter propofed to find the difference of
the direction of the two lines T′ P′, t p; (Figure 1). the firft
of which marks the longitude really obferved, the Earth being
at T′, and the Comet at C′; and the fecond, that which would
have been obferved, had the Earth been at t, and the Comet
at c.   The difference of thefe two directions, is the difference
of the two fmall angles, P′ T′ p, T′ p t; and is found in the
following manner.

§ 2. $\overset{34}{}$ The firft requifite is the fmall line T′ t, which, in the
middle of the Cord, is fenfibly the fame as the verfed fine of
half the Arch T T″, which is meafured by the Sun's motion in
longitude; and in any other part of the Cord, it may be taken
as proportional to the rectangle T t × t T″; that is, to the
product of the two times t and t'.   If therefore we call the
verfed fine of the half difference between the firft and laft longitudes of the Sun, V, and the line T′ t fought, V′; we

FIG. I. have this proportion; $\frac{1}{4} t''^2 : t\,t' :: v : v' = \frac{4\,t\,t'\,v}{t''^2}$. Which gives the value of $T't$, equal to V'.

$\overset{35}{}$

§ 3. The fines of the two angles fought, will then be found by the following proportions :

1. $SC'^2 : ST'^2 = 1 :: T't : C'c = \dfrac{T't}{SC'^2}$. (Chap. 1. § 7).

2. $SC' : SP' :: C'c = \dfrac{T't}{SC'^2} : P'p = \dfrac{SP' \times T't}{SC'^3}$.

3. $SP' : ST' = 1 :: \text{Sin. } ST'P' : \text{Sin. } SP'T' = \dfrac{\text{Sin. } ST'P'}{SP'}$.

4. $tp : Tt :: \text{Sin. } ST'p : \text{Sin. } T'pt = \dfrac{\text{Sin. } ST'p \times T't}{tp}$.

5. $T'p : P'p = \dfrac{SP' \times T't}{SC'^3} :: \text{Sin. } SP'T' = \dfrac{\text{Sin. } ST'P'}{SP'} :$

$\text{Sin. } P'T'p = \dfrac{T't \times \text{Sin. } ST'P'}{T'p \times SC'^3}$.

$\overset{36}{}$

§ 4. The fourth proportion, gives the fine of one of the two angles fought ; and the fifth, gives the fine of the other ; and as the value $T't$, is the fame in the expreffions of the two fines, and the values, Sin. $ST'P'$, and $T'p$, in the fifth, are very nearly the fame as Sin. $ST'p$, and tp, in the fourth ; the fine given in the fifth proportion, may be found by dividing the fourth, by SC'^3.

$\overset{37}{}$

§ 5. The reduction fought is therefore thus found. We have the angle $ST'P'$, which is the elongation of the Comet from the Sun in longitude ; that is, the difference of the fecond longitude of the Comet obferved, from the fecond longitude of the Sun given by the ephemeris or calculation ; this angle

may be taken for $S\,T'\,p$: we have alfo $T\,t = \frac{4\,l\,l\,'\eta}{l''^2} = v'$. FIG. 1.
Having therefore affumed the length of $T'\,P'$, the angle $T'\,p\,t$
may be found by the fourth proportion, as $T'\,P'$ may be fubftituted for t p. Conftruction, or a fimple trigonometrical calculation, which will be given hereafter, will give the diftance
from the Sun, $S\,C'$; whofe Logarithm, tripled, being taken
from the Logarithm of the fine of the Angle found by the proportion 4; the remainder will be the logarithmic fine of the
fecond Angle. If in the expreffion of proportion 4, we chufe
to employ t p, inftead of $T'\,P'$, it may be eafily found by con-
ftruction, by taking on the line $S\,P'$, the line $P'\,p$, $= \frac{S\,P' \times T'\,t}{S\,C'}$.
And this method will be the moft exact, particularly when the
intervals of obfervation are great; in which cafe the verfed
fines $T'\,t$, $P'\,p$, will not be fo fmall when compared with $T'\,P'$,
as to allow of our taking $T'\,P'$, for t p, without fenfible error.
Of this cafe more hereafter.

$\S\,\frac{38}{5}$. It is evident that the Angle $P'\,T'\,p$, will be greater or
lefs than $T'\,p\,t$, according as the diftance of the Comet from
the Sun $S\,C'$ is greater, or lefs, than the mean diftance of the
Earth from the Sun, confidered as unity; if thefe diftances are
equal, the Angles will be equal; and the reduction, which is
their difference, will be $= 0$. It will be $=$ to o when the fecond obfervation was made either in conjunction, or oppofition;
for then the Angle $S\,T'\,P'$ will be equal to o, or $= 180°$, in
which cafes its fine will be $= 0$, and the two angles $= 0$. In
pofitions near thefe, the reduction will be generally very fmall,
and may be neglected fafely. Thefe cafes are frequent, for if
the Comet is nearer to the Sun than the Earth, it is often feen
in conjunction, from its latitude; and if the Comet be further

G

Fig. 1. from the Sun, than the Earth; it is often feen in oppofition; and not feldom, at a diftance from the Sun, nearly equal to that of the Earth. If there are many obfervations of the Comet, thofe may be chofen or found by interpolation, which anfwer to one or other of thefe conditions; but this trouble is fcarce neceffary, the reduction being fo eafily found.

§ 6.$^{\underline{39}}$ The facility of the calculation is evident; for the firft Angle, in the proportion, (No. 4). in the fraction, $\dfrac{\text{Sin } 3\,\text{T}'\text{P}' \times \text{T}'t}{t\,p}$, the numerator is conftant; having therefore once found its Logarithm, it is only neceffary for each pofition to add to it the arithmetical complement of the Logarithm of t p; that is, of the diftance T′ P′, affumed; and for the fecond Angle, (in the proportion, No. 5). nothing is to be found but S C′, in order to add the arithmetical complement of its triple Logarithm; and this operation need only be performed for the firft and fecond pofitions. S C′, which is the only line to be found, will be obtained near enough by conftruction, as a fmall error in the quantities t p, S C′, will make a very flight change in quantities fmall of themfelves. This reduction, when the approach to the real diftance is near, will ferve for all the future operations.

§. 7.$^{\underline{40}}$ The formula of this reduction, may become extremely defective, in the cafe, (which is, however, very rare) of the Comet's very near approach to the Earth, in which cafe the divifor t p being too fmall, the value of the fine will become too great; and may even become abfurd, being greater than the Radius, which is = unity. But this happens to all methods which neglect quantities not infinitely but phyfically fmall: in

Mr. Cotes's differential formulæ for Trigonometry, the error is often infinite, when the fides or angles are of 90 degrees. Many circumftances will point out the cafe of the Comet's too near approach to the Earth, of which more hereafter; but one of the moft ftriking will always be, the great velocity of the Comet's apparent motion. In this cafe, the reduction may be found in the following manner.

§ 8. Having affumed a diftance T′ P′, we have in the triangle S T′ P′, the fides S T′, T′ P′, and the Angle S T′ P′. Hence the Angle T′ S P′, and the fide S P′ muft be found by calculation.

The fecond latitude l' = P′ T′ C′ is given, from which may be found P′ C′ = T′ P′ × tang: l'. In the triangle S P′ C′, right angled at P′; S P′, and P′ C′, are given; whence S C′, will be found. The proportion No. 2, (§ 3.) will give P′ p: taking T′ t, P′ p, from S T′, S p′, we fhall have S t, S p, which with the Angle t S p, will give the Angle S t p; whofe difference from the Angle S T′ P′, which is the fecond obferved elongation of the Comet, e'; will be the reduction fought.

§ 9. The trigonometrical part of the operation may be difpenfed with; and the Angle T′ S P′, and the diftances S P′, and S C′, found by conftruction; then having P′ p, by the formula; t p, may be drawn, and the Angle S t p, found by conftruction. But when the diftance T′ P′, is fmall; in which cafe this method is principally ufeful; the trigonometrical method is much the beft. A fmall error in the diftance S C′, will not fenfibly affect the value of the line P′ p, employed to find the fide S p, of the triangle S t p. When the diftance T′ P′,

FIG. 1. is not very fmall, the firft method given will be fufficiently accurate; as may be tried by comparing the refults of the two methods.

§ 10. The lines T′ t, and C′ c; are not exactly the effects of the gravity of the Earth, and Comet, towards the Sun : thefe effects would be expreffed by tangents drawn from the extremities of the arches T, and C; and through the other extremities T″, C″; lines parallel to the radii S T, S P. But in the orbit of the Earth, this line is almoft exactly double the verfed fine of the half arch, which we have employed; and in other points as t, the proportion given, T t × t T″, for the variation of the line T′ t; is almoft exact; the point S, being very near the centre of the Earth's orbit, which is very nearly circular. In the parabolic arch, the point S, being diftant from the centre of the ofculating circle, the error in the fuppofition is rather greater, in taking the verfed fine C′ c, in the middle of the cord, for an effect of gravity, correfpondent to that obtained for the Earth; and in other parts, taking the fame line as proportioned to C c × c C″; in order at all times to obtain T′ t, and C′ c, reciprocally proportional to the fquares of the diftances S T′, and S C′; but this difference will not be confiderable, unlefs from the Comet's near approach to the Sun, the parabolic arch of the orbit, fhould from the velocity of the Comet's motion, be very confiderable. But in this cafe a Comet is feldom vifible from its proximity to the Sun. In this cafe it would alfo be poffible to correct the value of the line P′ p, more accurately; but this work may be avoided by ufing obfervations made at fhorter intervals; which with a fmaller parabolic arch, will ftill leave a motion fufficient to be employed in the inveftigation of the orbit; or obfervations may be

Fig. 1.

employed, made at a greater elongation from the Sun. This too great proximity, will be indicated by the Comet's fmall elongation from the Sun, its bright light, and great apparent velocity; but it will be apparent from the firft trial made for its diftance, after having employed the reduction of the fecond longitude, as given in the preceding fections, and which cannot caufe any confiderable error in the inveftigation of the diftance.

H

CHAPTER V.

On the Proportion of the Three Curtate Diſtances of the Comet from the Earth.

Fig. 3. §. $\overset{44}{1}$. LET the Points T, t, T″; and P, p, P″; in Figure 3; be the ſame as in the Figure 1: Let the Line P″ D, be pa-rallel and equal to T T″; and let P D, be met in d, by p d, parallel to P″ D; and join T D, and T d. The Right Line P D, will then be the Orbit of the Comet, relative to its mo-tion in the Cord, if the Earth be conſidered as immoveable at T. The Line T D, will be parallel, and equal to T″ P″; becauſe they are bounded by T T″, and P″ D, which are equal, and parallel by conſtruction; and for the ſame reaſon T d, will be parallel and equal to t p: we alſo have T T″, T t; and P P″, P p; proportional to the times; whence we have the following proportions: D P″: d p ∷ P P″: P p ∷ T T″: T t; and as D P″ is = T T″, we ſhall alſo have d p = T t; that is, two equal and parallel lines, bounded by T d, t p.

§. $\overset{45}{2}$. Therefore T P, T d, T D, are the directions of the three longitudes of the Comet, the extreme ones as obſerved, and the middle one reduced to the Cord; and the Angles

P T d, d T D, P T D, will be the motions in longitude, anfwering to the times t, t', t''. Let thefe motions be called m,
m', m''; m'' being $= m + m'$; and we fhall have the following
proportions and' values:

1. $\dfrac{T P}{Sin.\ T d P} = \dfrac{P d}{Sin.\ P T d}$, and $\dfrac{T D}{Sin.\ T d D} = \dfrac{d D}{Sin.\ d T D}$; and becaufe the Sines of the two Supplements, T d D, T d P; are
equal and that T D $=$ T″ P″; we fay; T P : T″ P″::

$\dfrac{P d}{Sin.\ P T d} : \dfrac{d D}{Sin.\ d T D} :: \dfrac{t}{Sin.\ m} : \dfrac{t'}{Sin.\ m'}$.

2. $\dfrac{T d}{Sin.\ T D d} = \dfrac{D d}{Sin.\ D T d}$, and $\dfrac{T P}{Sin.\ T D P} = \dfrac{D P}{Sin.\ D T P}$: and

becaufe the Angle T D d, is the fame as T D P, and that

T d $=$ t p, we fay; t p : T P :: $\dfrac{D d}{Sin.\ D T d} : \dfrac{D P}{Sin.\ D T P} :: \dfrac{t'}{Sin.\ m'}$:

$\dfrac{t''}{Sin.\ m''}$

3. $\dfrac{T d}{Sin.\ T P d} = \dfrac{P d}{Sin.\ P T d}$; and $\dfrac{T D}{Sin.\ T P D} = \dfrac{D P}{Sin.\ D T P}$; and becaufe the Angle T P d, is the fame as T P D, and that T d
is $=$ t p, and T D $=$ T″ P″, we fay; t p : T″ P″:: $\dfrac{P d}{Sin.\ P T d}$:

$\dfrac{D P}{Sin.\ D T P} :: \dfrac{t}{Sin.\ m} : \dfrac{t''}{Sin.\ m''}$.

§ 3. By comparing thefe three proportions, it will be evident, that having taken at pleafure any two, out of the three
Curtate Diftances; one will be to the other, in a ratio compofed of the direct ratio, of the times elapfed between the obfervations taken, and the third; and the reciprocal ratio, of
the fines of the motions in longitude, anfwering to thofe times.

FIG. 3. Having therefore the times t, t', t''; and the motions in longitude m, m', m''; if any one diftance is known or affumed by the method of falfe pofition, the two others may be known; and as in each of the triangles, D T d, d T P, D T P; the Angle at T, is given, and by fubftituting T D, T d, for T''P'', t p, the ratio of the fides is obtained; the Angles at the bafe, and the ratio of the fides, to the bafe, may be found trigonometrically.

CHAPTER VI.

*Of the Graphical Delineation of the Orbit of the Earth,
and the Parabola of a Comet; and their Division into
Months and Days.*

FIG. 4.

§ $\frac{47}{1}$. THOUGH the Orbit of the Earth, be an Ellipsis; yet
the difference of the axes is so small, being only $\frac{14}{100000}$; that
it is quite insensible in construction; and an excentric circle,
will accurately represent it; care being taken to place it truly
in the Ecliptic.

§ $\frac{48}{2}$. Take therefore from a scale of equal parts, a radius of
about 1100 such parts, as the Earth's mean distance from the
Sun is meant to contain 1000, of; and strike a circle (Figure 4).
which represents the Ecliptic with the Sun S, in its centre. Set
off the Radius, on the circumference, all round; beginning at
pleasure any where; this will divide the circle into six parts of
two signs each; which may be subdivided where wanted, into
signs, and degrees. At 9. 10. (*a* Figure 4). draw a diameter
through the Ecliptic, which will be the Axis of the Earth's
Orbit. From the Sun S, towards the point *a*, set off ,017 from

I

FIG. 4. the fcale of equal parts; and from this point with a Radius of 1000, ftrike a circle for the Orbit of the Earth.

§ 3. $\overline{49}$ The places of the Earth in her Orbit will be then marked thus. Take from the Ephemeris the longitude of the Sun for the day wanted, and add fix figns ; the longitude of the Earth being ever directly oppofite to that of the Sun ; then lay a ruler from that point of the Ecliptic to the Sun S, and mark the Earth's Orbit, where the edge of the ruler cuts it.

FIG. 6. § 4. $\overline{50}$ The Orbit of a Comet whofe focal diftance and direc-trix are known, is thus drawn. Let S, (Figure 6). be the fo-cus, D D', the Directrix; from the focus, let fall a perpendicular on the Directrix at A ; biffect this line, as at V ; then V, will be the Vertex, and A V S, the Axis of the Parabola ; which prolong at pleafure from the focus, in an oppofite direction to the Vertex. Through the Axis, draw lines C, C, &c. parallel to the Directrix, and pretty near each other ; take in the com-paffes, the diftance A C, of each line C, from the Directrix, on the Axis ; and then fetting one foot of the compaffes in the Focus S, fet off that diftance to E, E, on the line from which the diftance was taken. Thefe points will be in the Parabola. For having drawn the lines E F, E F, from the points E, E ; perpendicular to the Directrix, they will be parallel and equal to C A, which by conftruction, is equal to S E, S E : There-fore E F, will be equal to S E ; which is the effential property of the Parabola.

Through the points thus found, the Parabolic Curve muft be drawn by hand ; but the part near the Vertex, which is the moft difficult to draw, may be done with the compaffes, by taking a

Radius, fomething greater than A S, as V K; then the Circular
Arch, from the centre K, will be found fenfibly to coincide
with feveral of the points near the Vertex, on each fide; and
with Radii rather greater, feveral more may be drawn through
in the fame manner, without fenfible error.

§ $\frac{54}{5}$. To divide the Parabola into days, we fhall make ufe of
the following moft elegant Theorem of Newton; invented and demonftrated by him in the firft book of the Principia, (Section 6,
Propofition 30.) but firft applied by the Abbè Bofcovich to this ufe.

While the Comet has an unequable motion in the Parabola;
if a circle be conceived to pafs through its place in its Orbit,
through the Sun, which is its focus; and through the Vertex of
the Orbit; the centre of this circle, will have an equable motion along a line perpendicular to the Axis, and biffecting the
perihelion diftance.

§ $\frac{54}{6}$. Let M V M'. (Figure 7, 8). be the Parabola defcribed
by a Comet whofe place is C; the Angle V S C, being acute in
the firft; and obtufe in the fecond figure. Draw B B', perpendicular to the Axis S V, and biffecting the perihelion diftance
at A, and let a circle pafs through the points V, S, C; its centre O, will be ever in the line B B', as is obvious on infpection;
and its motion along B B', will be uniform; from whence arifes
the great utility and elegance of the Theorem.

§ $\frac{55}{7}$. The demonftration of the rule is eafy. Draw the Ordinate C E, (Figure 7 and 8). and from O, draw O, F, parallel
to the Axis V S, and meeting C E, in the point F. Let V E
be $= x$; C E $= y$; A O $=$ E F $= z$; V A $=$ A S $= e$.

FIG. 7, 8. The parameter of the Axis $= 4\,S\,V = 8\,e$: and by the nature of the Parabola, $y^2 = 8\,e\,x$. By the difcoveries of Archimedes, the Parabolic Area $V\,E\,C$, is $= \frac{2}{3}\,V\,E \times E\,C = \frac{2}{3}\,x\,y$, and the Triangle $S\,E\,C = \frac{1}{2}\,E\,C \times E\,S = \frac{1}{2}\,y\,(\pm\,V\,S \mp\,V\,E)$ $= \pm\,e\,y \mp \frac{1}{2}\,x\,y$. Hence the Area of the Parabolic Sector $V\,S\,C = V\,E\,C \mp S\,E\,C = \frac{2}{3}\,x\,y + e\,y - \frac{1}{2}\,x\,y = \frac{1}{6}\,x\,y + e\,y$.

§ 8. We have alfo $O\,F = \pm\,A\,V \mp\,V\,E = \pm\,e \mp\,x$; $C\,F = C\,E - E\,F = y - z$; and as the Square, $O\,C^2$ is $= O\,V^2$, we have $O\,F^2 + C\,F^2 = A\,V^2 + A\,O^2$; that is; $e^2 - 2\,e\,x + x^2 + y^2 - 2\,y\,z + z^2 = e^2 + z^2$, or $x^2 - 2\,e\,x + y^2 - 2\,y\,z = 0$. But $- 2\,e\,x + y^2$, is $= - 2\,e\,x + 8\,e\,x = 6\,e\,x$; therefore $x^2 + 6\,e\,x = 2\,y\,z$; and $x^2\,y + 6\,e\,x\,y = 2\,y^2\,z = 16\,e\,x\,z$; that is $x\,y + 6\,e\,y = 16\,e\,z$; or $\frac{1}{6}\,x\,y + e\,y = \frac{8}{3}\,e\,z$. Now in this laft equation, the firft member is the value before found, for the Parabolic Sector; which is proportional to the time; therefore the fecond member is alfo proportional to the time; and as in this member, $\frac{8}{3}\,e$, is a conftant quantity, therefore z is proportional to the time; that is, the line A O, will vary as the times; therefore the motion of the point O, in the line A O, will be uniform.

§ 9. When the point C, (Figure 7). is extremely near the Vertex V; we fhall have V C, $= \frac{8}{3}\,A\,O$; that is the velocity of the Comet, at its perihelion; is to the velocity of the centre O; as 8, to 3.

§ 10. When the Angle V S C, (Figure 8). is a right Angle; the centre O, of the circle paffing through V, S, C, will biffect the hypothenuſe V C ; we ſhall therefore have A O = ½ C S = S V ; that is, the ſpace which the centre O moves through, while the Comet moves through 90° of Anomaly, is equal to the perihelion diſtance.

§ 11. Hence we deduce with great eaſe the diviſion of this line into days. We know by the common theory of the motion of Comets in parabolic Orbits, that when the perihelion diſtance is equal to the mean diſtance of the Earth from the Sun, the Comet moves 90° of Anomaly in about 109 days and $\frac{6}{10}$; (the Logarithm of the exact time is 2.0398717); and that in other parabolic orbits, the ſquares of the times for the ſame anomalies, are as the cubes of the perihelion diſtances ; which anſwer for all Parabolas (they being all ſimilar among themſelves), to the third law of Kepler for Elliptic Orbits. Adding therefore this conſtant Logarithm to the Logarithm of the perihelion diſtance, and half the latter Logarithm, (which anſwers to multiplying it by 3, and dividing it by 2), we have a Logarithm, whoſe number is the number of days employed by the centre O, to move the diſtance A O ; (Fig. 8.) equal to the perihelion diſtance of the Comet. The following is an example of this calculation applied to the Comet of 1769.

Logarithm of the perihelion diſtance —	9.089002
Half of it — — —	9.544501
Conſtant Logarithm — —	2.039872
Sum —	0.673375
Whoſe number is — —	4.7138

K

FIG. 7, 8. The number of days and decimals of a day, employed by the Comet, to move 90° of anomaly*.

This gives at once a fcale for the divifion of the Orbit, the perihelion diftance being known.

FIG. 7. § $\overset{60}{12}$. Having on the conftructed Orbit, a point C', (Fig. 7). deduced from obfervation; its point O', is eafily found, by joining the points S C', by a line, and biffecting it as in P. A perpendicular to the line S C', from the point P, will cut the line B B', in the point O', required. Or from the points S, and C', with the compaffes opened at pleafure; ftrike arches interfecting each other as at N, and N'; a line drawn through the points N, and N', will cut the line B, B', in the point O', required †.

* The above method of finding the motion of the Comet, can fcarce ever give directly, an exact number of days for the divifion of the fcale; but the line of lines on the Sector fupplies that at once. The example of the Comet of 1769 will fhew this. The number of days anfwering to the perihelion diftance is 4. 7138. Now to get the fcale of days without decimal parts; take the perihelion diftance as given by the fcale, in the compaffes; and open the Sector till 471 on the line of lines is = to the perihelion diftance; then 4. 5. 6. &c. on the even decimals of the line of lines, are the motions of the Comet for thofe numbers of days, on the line B B'.

† The demonftration of this operation is eafy. In the Triangles P C' O', and P S O', (Fig. 7). we have C'P = P S, by conftruction; the Angles C'P O', = S P O', by conftruction; they being Right Angles; and the fide P O', common. Therefore, the Triangles P C O', and P S O', are equal; and C' O', = S O'; which was required.

Having by the Rule in the laft §, the number of days anfwering to the perihelion diftance: a fimple proportion, or the line of lines on the Sector, will give the length of the line to be fet off from O', on the line B B', for any required Place of the Comet, either before, or after, the Time, when it was in the point of its Orbit C', anfwering to O', on the line B B'.

FIG. 7.

§ $\overset{61}{13}$. The fame rule, will find the time of the Comet's arrival at the perihelion ; for as V S, taken on the fcale of equal parts ; is to A O ; fo is the number of days anfwering to the anomaly of 90°, as found above ; to the days, and decimals of a day, the Comet is diftant from the perihelion, either before or after it.

Thefe decimals are reduced to hours, and minutes ; by multiplying them by 24, and 60 ; but conftruction cannot pretend to the precifion of minutes.

§ $\overset{62}{14}$. When the point O, is near the point A, the Circle and Parabola are fo nearly coincident, that the interfection cannot be determined accurately. The Arch V C, may then be fet off equal to $\frac{2}{3}$ of A O.

§ $\overset{63}{15}$. Having two obfervations, which give two points C, C′; we fhall have two points O, O′, from which the fcale might be at once taken for the divifion of B B′, into days ; for as O O′, is to A O ; fo is the time elapfed between the two obfervations ; to the time to be added, or fubtracted, to get the arrival at the perihelion. But as in this method the obfervations are not far apart, the line O O will be fhort ; as therefore the calculation is fo fhort, it will be better to ufe it, to get the greater interval S V ; which will determine the fcale, with more accuracy, befides that it will ferve as a verification of the conftruction ; as in the fcale fo determined, if the operation has been well performed, the line O O′, will anfwer on the fcale to the time between the obfervations.

CHAPTER VII.

Of the Numerical Quantities to be prepared for the Con-
ftruction or Computation of the Comet's Orbit.

§ 1. ⁶⁴ HAVING in the foregoing Chapter, given the methods of drawing, and dividing, the Orbit of the Earth, and a
known Comet; we fhall now proceed to the inveftigation of
the Orbit of an unknown Comet, from three obfervations.

The firft thing to be done, is to prepare all thofe numbers,
which will be wanted in the courfe of the operation; as feek-
ing each, as it comes to be wanted, would much retard the
graphical part of the bufinefs. Thefe numbers may with convenience be arranged in a table; as may be feen at the end of
this tract, in the compleat example of the whole procefs, both
numerical, and graphical, applied to the finding the Orbit
of the Comet of 1769.

§ 2. ⁶⁵ The following are the numbers to be prepared.

Firft. The true times of the three obfervations; expreffed
in days, hours, minutes, and feconds.

Second.   The three obſerved right aſcenſions, of the Comet.

Third.   The three obſerved Declinations, of the Comet.

Fourth.   The three longitudes of the Sun, at the times of the three obſervations.

Fifth.   The three diſtances of the Sun, from the Earth, at the times of the three obſervations. Theſe and the Sun's longitudes may be taken from the Ephemeris.

Sixth.   The mean time of each obſervation, found by applying the equation of time, to the true times given as above.

Seventh. The three longitudes of the Comet, deduced from the obſerved right aſcenſions, and declinations.

Eighth.   The three latitudes of the Comet, deduced from the ſame.

Theſe latitudes we ſhall call l, l', l''.

Ninth.   Three numbers, t, t', t''. Of theſe, t, is the interval between the firſt, and ſecond obſervation; in mean time, reduced to minutes; t', is the interval between the ſecond, and third obſervations, reduced as above; and t'', is the ſum of t, and t'.

L

Tenth.   Three numbers, m, m', m''. Of thefe, m, is the difference between the firft longitude of the Comet, and the fecond, in degrees, and minutes; m', is the difference between the fecond, and third; and m'', is the fum of m, and m'.

Eleventh.   Three numbers, e, e', e''; which are the three elongations of the Comet from the Sun, in longitude, exprefled in degrees, and minutes; and are found by fubtracting each longitude of the Sun, from the correfponding longitude of the Comet.   If the latter is the leaft of the two; 12 figns, or 360°; muft be added to it.

Twelfth.   The conftant number, a; found by adding to the conftant Logarithm, 0. 756499, twice the Logarithm of the number t'', that is, the Logarithm of t''^2

Thirteenth.   The number, v; which is the Logarithmic verfed fine, of half the terreftrial arch.   This arch is found, by fubtracting the firft longitude of the Sun, from the third; and the verfed fine v, by fubtracting the cofine of half the arch from Radius = 1.

Fourteenth. The number v'; $= \frac{4vtt'}{t''^2}$.   The numbers v, t, t', and t''; have been found above.

Fifteenth.   Three numbers, L, L', L''; of which L, is v', multiplied by the Sine of e'; L', is $= \frac{t'}{t'\, \text{Sine}\, m''}$, and L″, is $\dfrac{t''}{t\ \text{Sine}\ m''}$. L, is ufed in finding the reduction of the fecond longitude of the Comet; and L′, and L″, in finding the firft and third curtate diftances of the Comet from the Earth, from the fuppofed fecond curtate diftance, and from the numbers m, and m', corrected by the above-mentioned reduction.

CHAPTER VIII.

Determination of the Diſtances of the Comet from the Earth and Sun, by the Graphical Method.

FIG. 4. § 1. $\overset{72}{}$ HAVING thus prepared all numbers, to be uſed in the courſe of the conſtruction; we proceed to draw the Ecliptic, and Orbit of the Earth; (Figure 4.) according to the directions, given in Chapter 6. The Radius of the Earth's Orbit, ſhould not be leſs than ſix inches; but a greater ſize will be better, becauſe capable of greater accuracy.

§ 2. $\overset{73}{}$ By adding ſix ſigns, to the longitudes of the Sun; we ſhall have the three longitudes of the Earth, Q, Q′, and Q″. Laying a ruler, from each of them, to the Sun S; draw the lines S Q, S Q′, S Q″; cutting the Earth's Orbit, in the Earth's three Places, T, T′, T″.

§ 3. $\overset{74}{}$ From the Earth's three places T, T′, T″; draw the indefinite lines E, E′, E″; making the angles S T E, S T′ E′, S T″ E″; reſpectively equal, to the three elongations of the Comet from the Sun, e, e′, e″; taking care to place the Comet's elongation to the eaſt or weſt of the Sun, as the obſervations give it.

FIG. 4.

This may be done very well, by the line of cords on a good sector; or from the line of sines, taking double the sine of half the angle.

The elongations of the Comet, may also be thus drawn. Mark the observed longitude on the Ecliptic; and draw a line from it through the Sun. From the Earth's place in its orbit draw a line parallel to the former, which will be the elongation required. A good parallel ruler does this at once.

FIG. 5.

§ 4. On a paper apart, prepare a right angle, A T B, (Fig. 5.) and from the point, T, draw the lines, T F, T F', T F''; prolonged indefinitely, making the angles F T B, F' T B, F'' T B; equal to the three geocentric latitudes, of the Comet.

FIG. 4.

§ 5. Draw the Cord T, T''; (Fig. 4.) which will be intersected by the Radius S T, in a point t: then take a paper with a strait edge; (See Chap. 1. § 11.) and mark off the points, T, t, T''; which are represented as transferred on the line P, P''; in G, p, G'. Then according to the considerations, explained in Chap. 1. § 12. move the edge, with the points G p G', on it, till a judgment is formed of the position, and distance, to be first assumed, for the second curtate distance, T' P'. In addition to those considerations, we may observe, that the points, G, G', must be both within the lines, T E, and T'' E'', when the Cord of the Comet's motion, is greater than that of the Earth's; or both beyond those lines, when the Comet's Cord is less than the Earth's; and the parts of the line, G p G', cut off by the lines T E, and T'' E''; or the space on the line G G',

M

Fig. 4. prolonged to the lines T E, and T″ E″; muſt be proportional, to the parts of the line, G p, and p G′*.

The judgment of the diſtance to be fixed on, would be very eaſy, if the projected Cord of the Comet's Orbit P P″, was equal to the real Cord, C C″; for then, as has been ſaid before; it

---

* An example of this method of judging, will tend to elucidate the text. Let T, T′, T″; (Fig. 9.) be the places of the Earth; and E, E′, E″; the lines of elongation of the Comet; and the lines, 1, 1; 2, 2, &c. re-preſent the edge of the paper, with the T, p, T′; of the Cord of the Earth's motion; repreſented by G, p, G, applied to the lines, E, E′, E″; in different poſitions; it is firſt evident, that the line G, G′, muſt be ap-plied in the direction here given to it, in order to make the ſpaces inter-cepted between E, E′; and E′, E″, nearly proportional to the times. It is clear, that the poſition, 1, 1; is erroneous, the Comet at p, being within the Earth's orbit; and the ſpace intercepted by the lines E, E″, much leſs than the cord of the Earth's motion; G G′. The poſition 2, 2, is alſo wrong, the ſpace between E, and E″, being but little greater than G G′, though at p, the Comet is at the ſame diſtance from the Sun, with the Earth; in which caſe the cord of its motion, is to that of the Earth, as 14 to 10. Moving the line further off, as to 4, 4, the oppoſite error takes place; for now, though the place of the Comet at p, is conſiderably with-out the Earth's orbit; the cord of its motion, or the ſpace intercepted by E, and E″; is very much increaſed. In the poſition 3, 3, the conditions required are nearly anſwered; the point p, is a little without the Earth's orbit; and the ſpace intercepted by E, E″, is rather leſs than the propor-tion of 14 to 10, compared with G G′. The ſpace between G′, and the interſection of E″; is made rather greater than that between G, and the interſection of E; becauſe the velocity of the Comet's motion will increaſe, as it approaches towards the Sun; and p, is kept a little within the line E′, as directed. T′ P, is the curtate diſtance of the Comet from the Earth, as aſſumed by this firſt judgment.

would, at twice the Earth's diſtance from the Sun, be equal to the Earth's Cord; at an equal diſtance from the Sun, be to it as $\sqrt{2}$, to 1; or nearly as 14 to 10; and at half the Earth's diſtance from the Sun, it would be double the Earth's Cord.

But we muſt conſider that the obliquity of the Cord C C″, FIG. 1. to the plane of the Ecliptic, (Fig. 1.) makes it greater than the P P″, judged of from the ſtrait edge, with the points, G p G′; and a conjecture may be formed of the quantity of this obliquity, from the difference of the two elevations P C, P″ C″; (or their ſum, if the latitudes are on different ſides of the Ecliptic); which are the tangents of the obſerved latitudes, making the curtate diſtances T P, and T″ P″, Radii.

This difference of elevations, may be conſidered as the perpendicular of a right angled triangle, whoſe Hypothenuſe, is the real Cometary Cord, ſought for; and its baſe, is the line we are trying on the paper, with the edge G p G

This proportion, is not meant to be computed, or even conſtructed; but to guide the mind in this firſt eſtimation, which, with a little habit and attention, will generally be near the truth; and ſave much trouble in the future proceſs.

§ 6. Having now aſſumed a length for T′ P′, on the line FIG. 4, 5. T′ E′; we proceed to find the lengths of T P, and T″ P″, from it. This would be very ſhort, if the aſſumed T′ P′, and the ſecond longitude of the Comet, wanted no correction; which is the caſe, when the ſecond longitude is equal to that of the Sun, or ſix ſigns from it, or near theſe two poſitions. It then would be only neceſſary to apply the proportions given in Chapter 5. § 2. which are

$$\mathrm{T\,P} \;=\; \frac{t'' \; \mathrm{Sin.}\, m'}{t \; \mathrm{Sin.}\, m''} \times \mathrm{T'\,P'}$$

$$\text{and} \quad \mathrm{T''\,P''} \;=\; \frac{t'' \; \mathrm{Sin.}\, m}{t \; \mathrm{Sin.}\, m''} \times \mathrm{T'\,P'.}$$

Having before prepared the numbers L′, and L″; of which L′, is $= \dfrac{t''}{t' \; \mathrm{Sin.}\, m''}$, and L″, is $= \dfrac{t''}{t \; \mathrm{Sin.}\, m''}$; the operation will be

$$\mathrm{T\,P} \;=\; \mathrm{L'} \times \mathrm{Sin.}\, m' \times \mathrm{T'\,P'.}$$
$$\mathrm{T''\,P''} \;=\; \mathrm{L''} \times \mathrm{Sin.}\, m \times \mathrm{T'\,P'.}$$

§ $\frac{78}{7}$. As however the correction of the ſecond longitude is ſhort, always uſeful, and often neceſſary, it will be better always to make uſe of it; and we ſhall now determine this correction.

FIG. 5.   The aſſumed length of T′ P′, muſt be firſt ſet off on the line T B, (Fig. 5.) as from T to P′; and from P′, a perpendicular muſt be raiſed meeting the line, T E′, in the point C′.

FIG. 4, 5.   The diſtance S P′, muſt then be taken from Figure 4, and ſet off on the line T B, from P′ to S′, (Figure 5.) The diſtance S′ C′, meaſured from the point S′, on the line T B; to C′, on the line T F′; will be the Radius Vector of the Comet at the ſecond obſervation; and the line P′ C′, is its perpendicular diſtance from the plane of the Ecliptic. For it is evident that the lines P′ C′, and S′ C′, (Figure 5.) repreſent the P′ C′, and

S C', of Figure 1. For the triangle P' T C', (Figure 5.) right-angled at P', has the angle at T, prepared, equal to the geocentric latitude of the Comet obferved; and the fide T P', is the curtate diftance of the Comet from the Earth, transferred from Figure 4. This triangle, is therefore equivalent to the triangle T' P' C', (Figure 1.) and the fide P' C', is the tangent of the latitude, equal to the perpendicular diftance of the Comet from the plane of the Ecliptic. In like manner the triangle P' S' C', (Figure 5.) right-angled at P', has the fide S' C', transferred from Figure 4, which is the curtate Radius Vector; and the fide P' C'; proved before to be the perpendicular diftance of the Comet, from the plane of the Ecliptic; this triangle is therefore equivalent to the triangle S P' C', (Figure 1.) and S' C', is the Radius Vector of the Comet.

§ $\overset{80}{8}$. The next ftep is to get the value of P'p, and t p. The rule for finding P'p is given, Chapter 4, § 3; No. 2. the equation is $\frac{\text{S P}' \times \text{T}'\text{t}}{\text{S C}'^3}$. T't, is V', prepared before Chapter 7, § 2; and S P', and S C', were obtained graphically in the laft §. Having therefore folved this fhort equation, (fee the example,) and found the value of P p, in numbers; we fet it off from P', Fig. 4, to p, on the line S P' Having in like manner fet off V', which is T't, from T', to t, Figure 4, on the line S T'; we take the length t'p, and find its value on the fcale.

§ $\overset{79,\ 80}{9}$. Having thus got t p, we proceed to find the angles T'p t, and P'T'p. The rule for finding them is in Chap-

N

Fig. 4. ter 4, § 3. The fine of T′p t, is $\dfrac{\text{Sin. S T′p} \times \text{T′t}}{\text{t p}}$; and fubfti-

tuting the angle S T′ P′, which is the fecond elongation of the Comet, e', for S T′ p, we have before prepared L, Chapter 7, § 2. The equation therefore now becomes, Sin. T′p t $= \dfrac{\text{L}}{\text{t p}}$. (See the example.) The fine of P′T′ p, is equal to the fine of T p t, juft found, divided by S C′². The difference of thefe two angles, is the reduction fought, which we call y.

§ $\overset{80}{10}$. In fome very rare cafes, it will not be fafe to fubftitute the angle S T′P′, in the above formula, for S T′ p ; but then the angle S T′ p, may be found by conftruction. For having fet off P′ p, as before directed, draw the line T′ p; and the value of S T′ p, may be found by the line of fines on the Sector, and ufed in the equation inftead of S T′ P′. But in this cafe, which only can happen when the Comet is very near the Earth, and of courfe the T′ P′, very fhort ; it will be much the beft way to find the correction by the trigonometrical method given in Chapter 4, § 8.) and which is detailed in the example of the trigonometrical operation ; at leaft, when the pofitions come pretty near the truth.

§ $\overset{81}{11}$. Having thus found the reduction y, we muft by it correct the numbers m, and m'; (Chapter 1, § 6.) This is done, by adding it to one, and fubtracting it from the other, accord- ing to the following rule. If the line S C′, found in Figure 4, is greater than the Radius of the Earth's Orbit, if the motion of the Comet is direct, and T′ p, is contrary to the order of the figns, with refpect to T′ P′; the reduction muft be added to m, and fubtracted from m' : for by paffing from T′ P′, to T′ p, we diminiſh the longitude, and conſequently by paſſing from T′p, to t p, we increaſe it; and this increaſe is greater than the diminution; becauſe that the diviſion by S C′², will in this caſe, make the angle P′T′p, leſs than t p T′; therefore the ſecond longitude, will be increaſed by the reduction; and therefore the interval m, between it, and the firſt longitude, will be increaſed; and of courſe the interval m′, between it, and the third longitude, will be diminiſhed.—If any two, of the above three conditions, be changed, the correction muſt be applied the ſame way; but if one, or all the three conditions, be changed; then the reduction muſt be ſubtracted from m, and added to m′.

§ 12. Having thus corrected the numbers, m, and m′; we proceed to find the diſtances, T P, and T″P″. The equations for finding them, are $TP = \frac{t'' \text{ Sin. } m'}{t' \text{ Sin. } m''} \times t\,p$; and $T''P'' = \frac{t'' \text{ Sin. } m}{t \text{ Sin. } m''} \times t\,p$. We have before (Chapter 7, § 2.) prepared $L = \frac{t''}{t' \text{ Sin. } m''}$, and $L'' = \frac{t''}{t \text{ Sin. } m''}$; we have therefore, the equations $TP = t\,p \times \text{Sin. } m' \times L'$, and $T''P'' = t\,p \times \text{Sin. } m \times L''$.

Theſe equations will be eaſily ſolved, having L′, and L″. prepared before; t p, already uſed, and m, and m′, corrected. The values of theſe lines, muſt be taken on the ſcale, and ſet off on Figure 4, on the lines T E, T″ E″, to T P, T″ P″. In the following poſitions, the values t p, m, and m′, are the only ones to be changed, for the numbers L′, and L″, are conſtant.

§ 17. We now come to the compariſon of the equation, $b\,c^2 - \frac{c^4}{12\,b}$, with a. For this purpoſe we want the Radii Vec-

48

tores S C, S″ C″, of Figure 1, whofe fum is b; and the cord C C″ = c.

We find the Radii S C, and S C″, by the lines T P, T″ P″, found in the laft §, as S C′, was found from T′ P′. The lines T P, and T″ P″, muſt therefore be fet off on the line T B; (Figure 5.) as T P, T P″. The points, C, C″, in the lines T F, T F″, perpendicular to P, and P″, muſt be found. Then from Figure 4, we take the fpaces S P, and S P″; and fet them off in Figure 5, to P S, P″ S″: the intervals S C, and S″ C″, muſt then be taken; and their length, found on the fcale. Their fum is b.

§ $\overset{83}{18.}$ To get the length of C C″, the ſhorteſt of the two lines P C, P″ C″, in Figure 5, (which in this figure is P C,) muſt be fet off on the longeſt from C″, towards P″, if the latitudes are on the fame fide of the Ecliptic, (in this Figure, it reaches to I, on the line C″ P″,) or beyond C″, on the line C″ P″, pro-longed, if the latitudes are on oppoſite fides of the Ecliptic.

The line P P″, of Figure 4, muſt alfo be fet off from P″, to E. on the line T B. Then the line E I, will reprefent C C″, of Figure 1. For P″ E, P″ I, of Figure 5; are the C I, and C″ I, of Figure 1. The length of E I, taken on the fcale, is c. Twice the Logarithm of c, added to the Logarithm of b, gives the Logarithm of the firſt term, $b c^2$; whoſe number muſt be found; and the fum of the Logarithms of b, and of 12, fub tracted from four times the Logarithm of c, gives the Loga rithm of the fecond term, $\frac{c^4}{12\,b}$; whoſe number muſt be fub-tracted from the number of the firſt term, $b c^2$; to get the num ber, to be compared with a.

Fɪɢ. 4, 5.

§ 19. If thefe numbers are equal ; the pofition was right ; but that can hardly ever be the cafe. The difference of the number found, from a, muft therefore be taken, and called g ; with a pofitive or a negative fign, according as the number was greater, or lefs, than a.

§ 20. We muft now take a new pofition of the length of T′P′; which generally muft be taken greater, if the error g is negative ; and leffer, if g is pofitive. As the operation depends on the lengths of the Radii Vectores, S C, S C″; and of the Cord C C″: the change of T′P′ in the new pofition, muft be fuch, as will affect thofe lengths in a manner to correct the error. It will fometimes happen that increafing the length of T′P′, will fhorten the Radii Vectores : in this cafe if $b c^2 - \frac{c^4}{12 b}$ is lefs than a ; T′P′, muft be affumed lefs, to increafe the values of the equation ; but the leaft attention to the figure, will fhew whether that is the cafe or not. The change of T′P′ may be nearly judged of, from the quantity of the error, g ; the operations above defcribed, muft be repeated, as the fecond column of the table, in the example, fhews ; thefe will probably give a fecond error ; which we fhall call g'. b, is the difference of the length of T′P′, affumed in the two pofitions, which has a pofitive, or negative, fign ; according as the diftance in the fecond pofition, was increafed, or diminifhed. We muft then apply to the diftance taken for the fecond pofition, the value of the equation $\frac{g' b}{g - g'}$; and with this new affumed length, repeat the operation, till the error is very fmall ; but generally it will be very fmall at the third pofition.

O

§ 21. When the error is fmall, it is not neceffary to repeat the whole operation, by a new pofition; but the lengths of T P, T′ P′, T″ P″, S C, S C′, C C″, may be found, by calling the two laft errors g, and g'; the change of length of the line we are correcting, h; which is pofitive, if the laft pofition lengthened it; and negative, if it fhortened it; and applying the equation $\frac{g'h}{g-g'}$, to the length of the line, given by the laft pofition. Any other quantity, may alfo be corrected in this method; but thefe are the only ones neceffary, for finding the elements of the orbit by conftruction. This whole operation is drawn from the common rule of falfe pofition; where from the errors of two pofitions, the true quantity fought, is deduced; and this rule almoft always anfwers. But there are cafes in which being near a Maximum or Minimum of error, three pofitions muft be ufed, and fometimes the error cannot be totally made to vanifh. But this laft cafe never happens in the computation of the obfervations of a Comet which really exifts; and the rare cafe which demands three pofitions, will be perceived from the want of fuccefs in the ufe of two. Thefe reflections, and the neceffity of fubfidiary methods, apply equally to certain cafes, in all refearches, in which falfe pofitions are employed.

CHAPTER IX.

Determination of the Elements of the Orbit, from the determined Distances.

§ 1. THE elements of the orbit of a Comet, are six; and we shall determine them, in the following order.

First. The Place of the Node.

Second. The Inclination of the Orbit to the Plane of the Ecliptic.

Third. The Perihelion Distance.

Fourth. The Place of the Perihelion on the Orbit of the Comet.

Fifth. The Time of the Comet's Arrival at the Perihelion.

Sixth. The Direction of its Motion round the Sun.

§ 2. To find the line of Nodes, take the last determined lengths of T P, T″ P″, and P P″; and set them off on the line T B, (Figure 5.) as in T P, T P″, P″ E. From P, and P″, FIG. 4, 5.

FIG. 4, 5. raife perpendiculars meeting F, and F″, in C, and C″. On the line C″ P″, from the point C″, fet off C″ I, = C P; towards P″, if the latitudes are of the fame denomination; beyond C″, if they are of different ones. Draw I E; and C″ R, parallel to I E; meeting the line T B, in R. Set off P″ R, on Figure 4, on the line P P; prolonged as far as is neceffary.* Through S, and R, draw a line, which will be the line of Nodes; and will give their pofition, where it cuts the Ecliptic in N, and N′.

Infpection fhews which is the afcending Node; and the place of that, is given in the elements of the Comet's Orbit.

FIG. 1, 4, 5. §3. The demonftration of this conftruction is eafy. The lines, P″ I, P″ C″, P″ E; of Figure 5, are fimilar to the lines, I C″, P″ C″, I C; of Figure 1; for P I, Figure 5, and I C″, Figure 1, are the differences of the two obferved latitudes, P″ C″, P C; and P″ E, Figure 5, and I C, Figure 1, are equal to P″ P, Figure 4; the two P″ C″, P C, being the fame in Figure 5, and Figure 1. We have therefore thefe two proportions in Figure 5, P″ I : P″ C :: P″ E : P″ R; and in Figure 1, I C″ : P″ C″ :: I C : P″ R. The three firft terms of each being equal, the laft terms will alfo be equal. Having therefore taken the line P″ R, in Figure 4; equal to the line

* In the prefent cafe, as the line C″ P″, is fuppofed greater than C P; (Fig. 5.) P″ R, muft be fet off on the line P″ P, prolonged (Fig. 4.) from P″, towards and beyond P. Had the latitudes been of different denominations, and C″ P″, the greateft, the line P″ R, would likewife have been fet off from P″, towards P. But if C P, is greater than C″ P″; (Fig. 5.) then P R muft be fet off on P″ P; Fig. 4.) from P, towards or beyond P″.

P″R, (Figure 5), P″R, (Figure 4), is equal to P″R, (Figure
1); and the line S R, in the Figure 1, and Figure 4, will be the
fame; and in both, will reprefent the line of interfection of the
plane of the Comet's Orbit, with the plane of the Ecliptic;
that is, the line of Nodes.

§ $\frac{91}{4.}$ P R, may alfo be found by numbers thus: Take from
the fcale, the length of the lines C P, C″ P″, (Figure 5), and
P″ P, (Figure 4); and fay, as the difference of C P, and C″ P,
(or their fum if the latitudes are contrary,) is to C P; fo is P″ P,
to P R. This will be much the beft way of proceeding, if the
difference of C P, and C″ P″, be fmall; for in that cafe, the
line P R, will be very long: and if C P, and C″ P″, are equal;
their difference being = o, the point R, is infinitely diftant.
And in neither cafe, can R, be obtained by conftruction; but
in the latter cafe, N S N, muft be drawn parallel to P P″,
(Figure 4); and in the former cafe, a fourth proportional muft be
found to the lines, P C, P″ C″, (Figure 5), and S P″, (Figure 4).
Its value on the fcale, muft be fet off on S P″, to S Z; (Figure 4).
Draw P Z, and the line of Nodes will be a parallel to it,
paffing through the Sun.

For let us fuppofe the fame point Z, in Figure 1; we fhall
have the following proportions; P″ S: S Z:: P″ C″: P C::
P″ R: P R; the firft, being the proportion employed to find,
P Z; and the fecond, by fimilar triangles: it is by this, evident
that Z P, is parallel to S R; nothing therefore remains to be
done, but to draw through S, the line, N S N; parallel to P Z.

In this cafe, to get the point Z, in the Figure 4, more exactly, as it will not be far diftant from P″; the numerical value

P

of the lines P C, P″ C″, may be found as follows. Take from the scale, the lengths of T P, and T″ P″. To the Logarithm of T P, add the Logarithmic Tangent of *l*; and to the Logarithm of T″ P″, add the Logarithmic Tangent of *l*′; which will give the Logarithms of the lengths P C, P″ C″. These little calculations are so short, that they may be always used to determine these lines; the point R; the line of Nodes; and inclination of the Orbit; more accurately than construction can do it.

$\overline{89}$
§ 5. For the Inclination of the Orbit, draw P D, in Figure 4; perpendicular to the line of Nodes; and set it off on the line T B, (Figure 5), from P, towards T, as P D, and draw C D. The angle P D C, will be the inclination sought. For if in Figure 1, we conceive a plane, perpendicular to the line S R, and passing through C; its intersections C D, P D, with the two planes S R C, and S R P, will be perpendicular to the line S R; the angle in D, will be the angle of inclination of the two planes; and this angle will be the same as the angle D, Figure 5, for C P, is the same in the two figures; and P D, is also the same, being the same as in Figure 4.

If in the same manner, we draw the perpendicular P″ D″, (Figure 4), and set it off on Figure 5, to P″ D″; we shall have the angle of inclination at D″. The determination will be most accurate, by employing the longest of the lines P D, P″ D″; but it will be best to find the inclination from them both, and take a mean between the results.

$\overline{90}$
§ 6. This angle may also be found, by taking the length of P D, (Figure 4), and making it Radius on the line of Tangents, on the Sector; and the line P C, (Figure 5), will give the angle of inclination; or from the lines of Figure 4, by the following

proportion. P D : T P :: Tangent P T C = Tangent *l* : Tangent P D C = Tangent inclination ; this proportion depending on the line P C ; common to the two triangles, T P C, and D P C ; in Figure 1.

§ $\frac{93}{7}$. To find the remaining Elements of the Orbit, we muſt ſet off the two lines, D C, D″ C″, (Figure 5), on the two lines, D P, D″ P″, (Figure 4); from D, and D″, to C, and C″; from the centres C, and C″, with the Radii C S, and C″ S, ſtrike arches F, F′. Draw a line F F′, which is a Tangent to both theſe arches. From S, let fall a perpendicular S X, on the line F F′; then S X, will be the axis of the Orbit : and biſſecting S X, in V ; we have the Perihelion Diſtance S V.

§ $\frac{94}{8}$. For conceive the plane of the Orbit, Figure 1, S R C; to move on the line of Nodes, till it lies flat on the plane of the Ecliptic ; the Parabola will then be applied to the Ecliptic ; ſo that if the point C, repreſent any point in the Parabola, each right line D C, will coincide with the line P D, prolonged ; and as D C, D″ C″, (Figure 5), are the ſame as in Figure 1 ; the points C C″, (Figure 4), will be in the Parabola, applied to the Ecliptic. The lines C F, and C″ F′, are equal to S C, and S C″, being (as above) ſtruck with the Radii S C, and S C″; and they are perpendicular to the Tangent F F′; therefore by the properties of the Parabola ; F F′, is the directrix. S X, drawn from the focus S, perpendicular to the directrix, will be the axis ; and S V = half S X, is the diſtance from the focus, to the vertex of the Parabola ; that is the perihelion diſtance ; whoſe length is to be taken from the ſcale*.

* The method given above, is totally inapplicable, when the Perihelion diſtance is ſmall, and of courſe the obſervations muſt be made at a great

$\frac{95}{}$ § 9. The point, where the Axis, S V; cuts the Ecliptic, as x, (Figure 4); gives the Longitude of the Perihelion, on the plane of the Comet's Orbit; generally called fimply the Longitude of the Perihelion. If its Longitude on the Ecliptic, be required,

diftance from the Perihelion; as in fuch cafes, the Parabolic Cord C C", points almoft directly to the Sun; and the Arches F, and F, will almoft abfolutely coincide. Indeed in all cafes where the obfervations are at a great diftance from the Perihelion, the method will be inaccurate; and the Perihelion diftance had beft be found by the numerical method, given in Chapter 12, § 4. The following graphical method will however do very well, in all cafes whatever unlefs the Cord C C", fhould be very fmall indeed; and will for general ufe be preferable to the method given in the text. It is taken, though with fome variations, from the Cometographie de Pingre, Volume 2, Page 344; and is quoted by Mr. Pingre from Lambert. Its demonftration, and the demonftration of Bofcovich's numerical rule from it, are my own.

P R O B L E M.

The Focus; Two Radii; and the included Angle, or the Parabolic Cord, being given; to conftruct the Parabola:

Let S, (Figure 10.) be the Focus, S C, and S C"; the two Radii; of which S C", is the greater: C S C", the angle included, or difference of Anomaly of the Radii; and C C", the Parabolic Cord. On C C", as a diameter, defcribe the Semicircle, C A C". Take the difference of the Radii = S C" — S C; and fet it off from the end of the leffer Radius, on the Semicircle, that is, from C, to A. Draw C" A, prolonged at pleafure, as to H. From the Focus S, let fall S H, perpendicular to C" H, and of courfe, parallel to C A; (the angle C A C", being a right angle as contained in a Semicircle,) then S H, will be the Axis of the Parabola. From H, on S H, prolonged beyond S; fet off the greater Radius, = S C", to D. Biffect S D, in V; then V, is the Vertex of the Parabola;

it is found as follows. Draw T G, (Figure 5), parallel to D C, or D" C"; in Figure 4, draw V V', from the Vertex of the Parabola, perpendicular to the line of Nodes, S R. Set off V V', on the line T G, (Figure 5), to T V; and draw V L, perpendicular to T B. Take T L, and fet it off on the line

S V, is the Perihelion Diftance; and a line drawn through D, perpendicular to the line D H, will be the Directrix.

The demonftration is eafy. In the Parabola, V C C", (Figure 10), let S, be the Focus; D H, the Axis; D M, the Directrix; and H C", an ordinate to the point C": and from C, and C", draw M A, and M" C", at Right Angles to the Directrix. Now we have by conftruction, M A = M" C" = D H; and by the effential property of the Parabola, M C = S C; and M" C" = S C". Therefore M A = S C" — S C. And the Angle C A C" is a Right Angle by conftruction. Therefore, if on the Cord C C", a Triangle be conftructed, as C A C"; having the Angle A, oppofite the Cord, a Right Angle, and the fide C A, adjacent to the leffer Radius Vector S C, equal to the difference of the Radii; then will the fide C A, be parallel to the Axis of the Parabola; and the fide A C", parallel to the Directrix. But this was done, in the conftruction of the Problem; for the Angle in a Semicircle is a Right Angle; and the fide C A was taken equal to the difference of the Radii. The reft follows of courfe.

When the Cord C C", is fmall; a greater accuracy in the conftruction may be obtained by prolonging the Cord at pleafure beyond the greater Radius, and taking any multiple of the Cord, as a Diameter for the Semicircle; then fetting off from C, an equimultiple of the difference of the Radii, on the Semicircle, a parallel to that line paffing through the Focus will, as before, be the Axis of the Parabola. Thus in Figure 10, C K, is taken = twice C C", on the Cord prolonged; and the Semicircle C L K is ftruck; on which C L = twice the difference of the Radii, is fet off, from C, to L; and C L, is a parallel to the Axis, as C A, was before.

Q

FIG. 4, 5.   V V', (Figure 4), from V' to u; through the points S u, draw a line cutting the Ecliptic in x'. x' is the longitude of the Perihelion in the Ecliptic. For if in Figure 1, we conceive the line V u, perpendicular to the Plane of the Ecliptic; drawn from the Vertex V, of the Parabola, C" C' C V; and the Plane V u V, perpendicular to the line of Nodes, at V'; the Angle V V' u, will be the inclination of the Orbit; and therefore equal to the Angle V T L, (Figure 5); in the Figure 5, T V, is equal to V V', (Figure 4); that is, to V V', (Figure 1); therefore V' u, in Figure 4, which is equal to T L, (Figure 5), will be equal to V u, (Figure 1); and S u, (Figure 4), to S u, (Figure 1); which gives the direction of the longitude of the Perihelion; therefore the direction of the line S u, (Figure 4), gives the longitude of the Perihelion in x'.

This Problem, is in fact the same operation by lines; as the Theorem for finding the Anomaly, given in Chapter 12, § 4; and that Theorem, may be demonstrated from it, thus.

Draw B C", parallel to C A; and B C, parallel to C" A. Then by Alternate Angles, the Angle B C" S, will be = C" S H; the Supplement of the Anomaly of the greater Radius. Now the Angle B C" S, is composed of the Angle C C" S; which is the Angle opposite to the lesser Radius, in the Triangle C S C"; and the Angle B C" C; which in the Right-Angled Triangle, B C" C; is thus found. As C C", to Radius; so is B C", to the Sine B C C", = Cosine B C" C. Therefore $\frac{B C''}{C C''}$ = Cosine B C" C. But C C", is the Cord of the Comet's motion in the Triangle C S C"; and B C", is by construction = C A = S C" — S C. Therefore, the Angle opposite the lesser Radius; + the Angle, whose Cosine, is the Cord divided by the difference of the Radii; is = the Supplement of the Anomaly of the Greater Radius; Q. E. D.

§ 10. $\overline{\underset{96}{}}$ The time of the Comet's paſſage by the Perihelion, is Fɪɢ. 4, 5. found by the method given in Chapter 6, § 13, &c.

§ 11. $\overline{\underset{99}{}}$ We have thus obtained the five firſt elements of the Comet's Orbit; and a bare inſpection of the Figure, gives the direction of its motion round the ſun, with reſpect to the order of the ſigns in the Ecliptic. This therefore is the Sixth Element ſought; and completes the Theory of the Comet's Orbit.

CHAPTER X.

Determination of the Place of the Comet, for any given Time.

§ 1. $\overline{\underset{100}{}}$ **I**N order to determine the Place of the Comet, either as feen from the Earth, or the Sun, for any given time; the Orbit of the Earth muft be drawn as directed in Chapter 6; and alfo the Orbit of the Comet, in its true dimenfions and pofition in the Ecliptic, according to the Elements determined in the laft Chapter. The Orbit is fuppofed to revolve on the line of Nodes, till its Plane coincides with the Plane of the Ecliptic. This we fhall call the applied Orbit of the Comet; to diftinguifh it from the projected Orbit, to be defcribed hereafter. (We may here obferve that the projected Orbit is likewife a Parabola). The Orbit muft be divided into days, according to the method given in Chapter 6, § 4 and 11.

§ 2. $\overline{\underset{100}{}}$ The line of Nodes N N', muft be alfo drawn through the Sun S, cutting the Ecliptic in the point of the Longitude of the afcending Node ; and prolonged each way at pleafure.

FIG. 1, 4, 5. § 3. $\overline{\underset{103}{}}$ The Places of the Comet in its projected Orbit, are then found as follows. From the Place of the Comet in its

applied Orbit as C''', (Figure 4), draw a line C''' D''', perpendicular to the line of Nodes N N'; and having prepared an
Angle as G T H, (Figure 5), equal to the inclination of the
Comet's Orbit; on the line T G, set off C''' D''', from Figure 4,
as to T H. From H, let fall a perpendicular on the line T B,
to M, and set off T M, in Figure 4, on the line C''' D''', from
D''', towards C''', as D''' P'''. Then P''', will be the Comet's
Place in its Orbit, projected on the Plane of the Ecliptic; and
the line H M, (Figure 5), will be the perpendicular distance
of the Comet from the Plane of the Ecliptic; which will be of
use in future operations.

It is evident that P''', is the projected Place of C'''; for the
Angle H T M, (Figure 5), is equal to the inclination of the
Orbit; and therefore the Triangle H M T, answers to the Triangle C D P, (Figure 1); and the line T M, to P D; in the
same figure. The line C''' D''', (Figure 4), is to P''' D'''; as
Radius, to the Cosine of the inclination of the Comet's Orbit:
and this Ratio is the same for any point that may be taken in
the Orbit*.

* For let Figure 11, be supposed a section of the Comet's Orbit, by a
Plane perpendicular to the Line of Nodes, and of course perpendicular to
the Planes of the Comet's Orbit, and the Ecliptic; then the Line T D,
will represent the Section of the Plane of the Ecliptic; C D, the Section
of the Comet's Orbit; and the Angle C D T, the Inclination of the Co-
met's Orbit. From the Centre D, with the Radius C D, describe the
Quadrant T C E, and from C, let fall the perpendicular C P, on the
Line T D, and draw C F, parallel to T D; then C P, is the Sine of the
Inclination of the Orbit, and C F = P D, the Cosine of the Inclination.
But P D, is the projected Place of the Comet, as directed to be drawn:

R

Fig. 4.

¹⁰³
§ 4. If it is not thought worth while to divide and project the whole Orbit of the Comet, as above directed; any point therein may be found by the same rule, when the Parabola is drawn. For the Perihelion Distance furnishes a Scale, from which a simple proportion will give the length of any line to be set off on B B′, for the point O, from whence the point C, is to be found in the applied Orbit, with the Radius S O. And from the point C, the last § shews how its projected point P, is to be obtained.

¹⁰⁴
§ 5. Having the Place of the Comet in its applied and projected Orbit; its Geocentric Longitude, and Latitude, are thus found. By the method given, Chapter 6, § 3, find the Place of the Earth in her Orbit; from the Place of the Earth, to the Place of the Comet in its projected Orbit, draw the line, (Figure 4), T‴ P‴: and parallel to T‴ P‴, draw a line passing through the Sun, as S Y; which will cut the Ecliptic in a point Y, which is the Geocentric Longitude of the Comet*.

therefore P‴ D‴, is the Cosine of the Inclination to the Radius C‴ D‴ C P, is the perpendicular Distance of the Comet, from the Plane of the Ecliptic; which therefore is the Sine of the Inclination of the Comet's Orbit, to the Radius C‴ D‴. The Sector therefore furnishes a much more expeditious, as well as accurate method, of obtaining both these Lines, than that given in the Text. For taking C‴ D‴, as Radius, on the Line of Sines; then the Cosine of the Inclination (= the Sine of the Coinclination) will give P‴ D‴; and the Sine of the Inclination, is the Perpendicular Distance of the Comet from the Plane of the Ecliptic.

* Or the Angle P‴ T‴ S, which is the Elongation of the Comet, from the Sun, may be measured by the Protractor or Sector; and this Angle,

§ 6. For the Geocentric Latitude; on the line T B, (Figure 5), from the point M, set off the line T''' P''' of Figure 4, to M Q, and draw Q H. The Angle M Q H, will be the Latitude sought. For H T M, being by construction the Angle of Inclination of the Orbit; the Triangle H M T will be similar to C P D, (Figure 1), and H M Q, to C P T, (Figure 1)*.

§ 7. The true Distance of the Comet from the Earth is Q H, (Figure 5), which answers to the Hypothenuse T C, of the Triangle C P T, (Figure 1). The distance from the Sun will be S C''', (Figure 4); the Heliocentric Longitude, will be determined by a line drawn from S, through P''', cutting the Eliptic in y; and the Heliocentric Latitude will be found by setting off S P''', (Figure 4), on the line T B, (Figure 5), from M, as to q. Draw q H, and the Angle M q H will be the Heliocentric Latitude of the Comet.

added to or subtracted from the Longitude of the Sun; according as the Elongation is West, or East of the Sun, will give the Geocentric Longitude of the Comet.

* This is done with more expedition as well as accuracy by the Sector; thus. Make T'' P''', Radius on the Line of Tangents; then H M, found before, § 2; will give the Angle of the Geocentric Latitude on the Line of Tangents.

In the same manner, with the Radius S P''', will the Heliocentric Latitude be found.

CHAPTER XI.

Determination of the Distances of the Comet from the Earth and Sun, by Trigonometrical Calculation.

FIG. 1, 4, 5. §¯147¯ 1. H AVING in the three last Chapters, given the method of finding the Elements of a Comet's Orbit by construction; we shall now proceed to the Trigonometrical method of finding them. This is susceptible of greater accuracy, but requires more labour and time; it is therefore always better to begin by the Graphical method, and having attained to a near approach to truth in this way, make use of Trigonometry, in case greater precision is required. In all these calculations, Seconds may be neglected, and a minute added when the Seconds are above 30; for the observations of Comets can scarce ever be certain to a minute, and very often err two.

§¯148¯ 2. If the reduction of the Second Longitude by the Graphical method is not thought sufficiently accurate; the Trigonometrical method given in Chapter 4, § 8, may be used; and having thus reduced the second Longitude, and corrected m, and m'; Chapter 8, § 11; and having L', and L"; Chapter 7, § 2; we obtain as in Chapter 8, § 6, and 12; the value T P, = L' × Sin. m' × t p; and T" P", = L" × Sin. m × t p.

§ 3. But if the reduction already found, be ufed; we take FIG. 1, 4, 5. the laft found value of T P, and having m, and m', already corrected; we get $T''P'', = \dfrac{t' \text{ Sin. } m}{t \text{ Sin. } m'} \times$ T P. We have alfo P C, = T P × Tang: l, and $P''C'', = T''P'' \times$ Tang: l''. We fhall then have Six Triangles to folve; three of which are oblique-angled, T S P, T″ S P″, P″ S P; and three are right-angled, S P C, S P″ C″, C I C″.

§ 4. In the Triangle T S P, we have given, the Side S T, = the firft diftance of the Earth from the Sun; T P, already found; and the Angle S T P, = e, the firft Elongation of the Comet from the Sun; from thefe we obtain the Angle T S P, and the Side S P, by the following Theorems.

Given: an Angle of a Triangle, with the two Sides containing it. Required: either of the other Angles, and the third Side.

For the Angles. As the fum of the Sides given,: to their difference:: fo is the Tangent of the half fum of the unknown Angles (found by fubtracting the given Angle from 180°): to the Tangent of an Angle, which added to the half fum, gives the Angle oppofite to the greater Side; and fubtracted from the half Sum, gives the Angle oppofite the leffer Side. In Logarithms thus:

a : c : Log: $\overline{\text{S T} + \text{T P}}$ + Log: $\overline{\text{S T} - \text{T P}}$ + Log: Tang: ½ Sum of two unknown Angles, = Log: Tang: of an Angle, to be added to, or fubtracted from the ½ Sum, to give the two unknown Angles.

S

For the Third Side. As the Sine of either Angle juſt found, : to the Side oppoſite : : ſo is the Sine of the given Angle, : to the Side required. In Logarithms thus :

$$a : c : Log : Sin : TSP + Log : TP + Log : Sin : STP = Log : SP.$$

§ 5. [150] In the Triangle T″ S P″; we have S T″, the third diſtance of the Earth from the Sun; T″ P″, already found; and the Angle S T″ P″, = e″, the third elongation; from which we obtain the Angle T″ S P″, and the Side S P″, as in the preceding §, we obtained T S P, and S P.

The difference of the Sun's firſt and third Longitude, is the Angle T S T″, which (in the caſe expreſſed in the Figure), ſubtracted from the Angle T″ S P″, leaves the Angle T S P″, and this ſubtracted from the Angle T S P, leaves the Angle P″ S P. In ſome caſes, addition muſt be employed inſtead of ſubtraction, to get this Angle; but it will always be obtained from the three Angles T S T″, T S P, T″ S P″; and a rough drawing will be a guide ſufficient to find when addition is to be employed, and when ſubtraction.

§ 6. [150] Having the Angle P S P″, with the Sides S P, S P″; we find the Side P P″, in the third Triangle P S P″; as S P was found in § 4.

§ 7. [151] Then in the right-angled Triangle S P C; having the Sides S P, P C; we obtain the Hypothenuſe S C; by the following Theorem.

Given the two Sides containing the Right Angle. Required the Hypothenuse.

First. As one Side : to the other Side :: So is Radius : to the Tangent of the Angle opposite the Second Side. In Logarithms thus :

$a : c :$ Log : $SP +$ Log. $PC =$ Log : Tang : PSC.

Then. As the Sine of the Angle just found : to the Side opposite to it :: So is Radius : to the Hypothenuse. In Logarithms :

$a : c :$ Log : Sin : $PSC +$ Log : $PC =$ Log : SC.

§ 8. $\overset{151}{}$ In the second Triangle $SP''C''$; the Sides SP'', $P''C''$; give SC''; as SC was found in the preceding §.

§ 9. $\overset{151}{}$ In the last Triangle CIC''; having the Sides, $CI = PP''$; $C''I = P''C'' \pm PC$; (the sign $+$ being used when the Latitudes are of contrary denominations), we obtain the Hypothenuse CC''; as SC was before found.

§ 10. $\overset{152}{}$ Thus we have the value of $SC + SC'' = b$; and $CC'' = c$; and having the value of a, prepared for the former process ; we compare the equation $bc^2 - \dfrac{c}{12b}$, with it, as before directed, Chapter 8, § 19 and 20, to find the error of the first position. Another position chosen from the quantity and quality of this first error, as in the former method, will give a second error ; and if the method of construction had been previously used to get an approach to the truth ; the errors will now be so small, that we may at once employ them as in Chapter 8, §. 21 ; to find the Distances and Cords, to be used in the determination of the Orbit.

CHAPTER XII.

*Determination of the Elements of the Comet's Orbit,
by Trigonometrical Calculation.*

FIG. 4, 5. § $\overset{154}{1}$. $\mathbf{H}$AVING obtained P C, P C'', and P P''; to a fufficient degree of accuracy; we proceed to fay as in Chapter 9,
§ 4, C'' I = P'' C'' $\pm$ P C : P C :: P P'' = C I : P R. The
Sign +, being ufed, when the two Latitudes are of contrary
denominations; and —, when of the fame.

This proportion gives P R. We have alfo S P found, Chapter 11, § 5; and when we refolved the Triangle P S P'', to
find P P'', Chapter 11, § 6; we alfo found the Angle S P'' P,
= S P'' R. We therefore find the Angle P'' S R, by the Rule
in Chapter 11, § 4; and having before, Chapter 11, § 5, found
the Angle T'' S P''; the fum of thefe two Angles (in the cafe
expreffed in the Figure) fubtracted from the third Heliocentric
Longitude of the Earth, expreffed by the Radius S T''; will
give the Longitude of the Node N, marked by the Line S R.

§ $\overset{154}{2}$. From the Angle P'' S R, and S P''; the Inclination of
the Orbit is alfo found by the following proportions:

1ft. $SP'' \times Sin. P''SR = P''D''$; $=$ a perpendicular let fall from the Comet's Curtate Place at P'', on the Line of Nodes.

2d. $\frac{P''D''}{P''C''} =$ Cotangent of the Inclination: $P''C''$, being the perpendicular Diftance of the Comet from the Plane of the Ecliptic*.

Therefore at once, fubftituting SP'', $\times$ Sin. $P''SR$, for $P''D''$; we fay; Cotangent of Inclination $= \frac{SP'' \times Sin. P'SR}{P''C''}$.

§ 3. $\overset{155}{}$ If the Latitudes are of contrary denominations, the point R, will fall between the points P, and P''; if they are of the fame denomination, it will fall on the Line PP'', prolonged beyond P, or P'', according as $P''C''$, is greater or lefs than PC; and to have the Longitude of the Line SR; if SR, and ST, are both on the fame fide of the Line SP''; the difference of the two Angles TSP'', $P''SR$, muft be taken, and fubtracted from the Longitude of ST, when the Angle $T''SP''$, is in a direction contrary to the order of the figns; and added, when it is according to that order. If the fum or difference to be fubtracted, is greater than that from which it is to be fubtracted; 360° muft be added; and when in additions the fum is above

---

* For in Fig. 11, PD, is the $P''D''$ of Fig. 4, and PC, the $P''C''$ of Fig. 5; now to the Radius PC, PD isTangent to the Angle PCD; which is the Coinclination, $=$ Cotangent of the Inclination. Therefore to find the Angle of Inclination PDC, we fay; as PC: Radius :: PD: Cotangent of Inclination; $= \frac{PD}{PC}$.

T

FIG. 12.

360°, 360 muſt be thrown off, as is always practiſed in Aſtronomical calculations. The Figure will be a guide to the calculation as to the additions and ſubtractions; and it is obvious which of the two, is the aſcending Node.

§ 4. The Perihelion Diſtance, and its Place on the Orbit, will be found by means of the Anomaly C″ S V, of the greater Radius S C″; (Figure 12), and this will be found as follows: Firſt, in the Triangle C S C″; we have the three Sides S C, S C″, C C″; the following Theorem will therefore give the Angle C C″ S:

Given: the three ſides of a Triangle, to find an Angle. Rule: From the half ſum of the three ſides, ſubtract each of the ſides, S C″ and C C″, containing the Angle ſought, C C″ S. Then in Logarithms we have

Sine of ½ C C″ S = ½ (+ Log: $\overline{\frac{1}{2}\text{ Sum} - C C''}$ + a : c : Log: S C″ + a : c : Log: C C″).

Then for the Angle B C″ C; draw C B from C, perpendicular to C″ F, and we have B C″ = the difference of C F and C″ F′; that is, of the two Radii Vectores S C and S C″. Then in the Triangle C C″ B, right-angled at B; we ſay; as C C″ : C″ B :: Rad: Sin. B C C″ = Cos: B C″ C. Or in Logarithms, Cos: B C″ C = Log: B C″ — Log: C C″. The ſum of the Angles C C″ S, and B C″ C, is B C″ S: and as F′ C″ is parallel to X S, the Angle B C″ S is the complement of the Angle V S C″, the Anomaly of the Greater Radius S C″.

We have therefore the following Rule applicable to every caſe:

In the Triangle formed by the two Radii Vectores and the Cord; find the Angle oppofite to the leffer Radius, and the Angle whofe Cofine is the difference of the two Radii divided by the Cord. The fupplement of their fum will be the Anomaly of the greater Radius. (See the Note on Chap. 9, § 8).

§. 5. We fhall then find the Perihelion Diftance by the following Theorem.

The Perihelion Diftance is equal to any Radius Vector, multiplied by the fquare of the Cofine of half its Anomaly. Therefore twice the Logarithmic Cofine of half the Anomaly found in the laft §, added to the Logarithm of the longer Radius, will give the Logarithm of the Perihelion diftance fought.

§ 6. The place of the Perihelion may be deduced from the fame Anomaly. We have already (Figure 1), the Angle P"SR; and its Tangent is to the Tangent of C"SR; as D"P", is to D"C"; for D"P", and D"C" are the Tangents of the Angles D"SP", D"SC" to the common Radius SD'; and as D"P", is to D"C"; fo is the Cofine of the Inclination to Radius. We therefore have this proportion: Tangent $C''SR = \dfrac{\text{Tangent } P''SR}{\text{Cofine Inclination}}$.

We have the Anomaly C"SV, confequently we may find V S N, which will be the diftance from the Perihelion to the Node on the Orbit.

§ 7. Having the Longitude of the Node N, and the diftance of the Perihelion from the Node; we have the longitude of the Perihelion in the Orbit. A rough drawn Figure will fhew whether fubtraction or addition muft be employed to find the Angle V S R, by the two Angles C"SR, C"SV; and the place of the Perihelion by the Node N, and the Angle V S N; ſuch a Figure will alſo ſhew the poſitions of all the Lines which include the Angles employed in this calculation, and be a guide in all caſes that may ariſe as to the different applications of the Theorems. It muſt be ever remembered that if the Perihelion V, found by the Anomaly C″ S V, does not fall on the ſhorteſt of the Radii Vectores, S C, S C″; it will be nearer the ſhorteſt than the longeſt of theſe Radii.

160

§ 8. The Perihelion diſtance and Anomaly already found, give the time of the arrival at the Perihelion. This is eaſily found from the Tables of the Parabola firſt computed by Halley, and univerſally uſed, which are given in an improved form, at the end of this work, or by the other Table of the Parabolic Fall of Comets, there given; the mode of uſing them is given at large, in the explanation and uſe of the Tables.

162

§ 9. Having thus obtained the five firſt Elements of the Comet's Orbit; the Figure before uſed in the Graphical Proceſs, will ſhew the direction of the Comet's motion, with reſpect to the order of the Sine; that is, whether it is direct, or retrograde.

CHAPTER XIII.

Determination of the Place of the Comet, for any other
given Time, by Calculation.

§ 1. $\overset{173}{}$ To find the Place of the Comet for another given Fig. 13.
time from the Elements already determined, the Tables of Co-
mets Motion already mentioned muft be ufed. From them
the Anomaly of the Comet muft be found for the given time;
and from it, the Radius Vector, which is equal to the Perihelion
Diftance divided by the Square of the Cofine of half the Anomaly. This is eafily done by Logarithms, being the fum of
the arithmetical complement of twice the Logarithmic Cofine
of the half Anomaly, and the Logarithm of the Perihelion
Diftance.

§ 2. $\overset{174}{}$ The reft of the operation for Comets is the fame as for
Planets. We have the diftance from the Node to the Peri-
helion, which is the Arch N V, Figure 13; we have found the
Anomaly, which is the Arch V C; and if the Comet's motion
is direct, and the time follows the arrival at the Perihelion; or
both thefe conditions are changed, V C, will follow the order of
the Signs; otherwife it will not. By the two Arches N V, V C,
with their direction, we obtain the Arch N C, which is the Co-

U

FIG. 13. met's diſtance from the Node; which with the Angle N, which is the Inclination of the Orbit, will give the Arch C P, the Comet's Heliocentric Latitude; and N P, which is the diſtance from the Node N, to the place of the Comet on the Ecliptic; and the Longitude of the Node being known, the Comet's Heliocentric Longitude is found from it. The Theorem for the Latitude C P, is; Sin. N C × Sin. N = Sin. C P. And for the Arch N P, the Theorem is, Tangent N C × Coſine N = Tangent N P.

§ 3. This might have been done from Figure 1, by Plane Trigonometry only; but the method above given is very ſimple, and generally uſed for the Planets. Transferring theſe Elements to Figure 1, we have the direction of the Line S P, which is the Heliocentric Longitude; the Radius S C; the Heliocentric Latitude C S P: from whence may be deduced the Curtate Diſtance S P, = S C × Coſine C S P. From the Ephemeris we find the Earth's place and diſtance from the Sun. The difference of the Heliocentric Longitude of the Comet, and the Earth, is the Angle T S P; which with the Sides S T, S P, will give the Angle S T P; and the Comet's curtate diſtance from the Earth, T P. The Angle S T P, is the elongation of the Comet from the Sun; from which, and the Sun's Longitude, the Comet's Geocentric Longitude is found: and the Geocentric Latitude is thus found; $\dfrac{S\,P \times Tang.\,P\,S\,C}{T\,P}$ = Tangent of the Geocentric Latitude.

§ 4. The whole of this long proceſs muſt be renewed for every place of the Comet which is wiſhed to be exactly found; whereas the method before given by conſtruction, Chapter 10 is very fhort and eafy; and fully exact enough to find where to look for the Comet after a few nights interval, or even to difcover any confiderable error in the firft determined Orbit, by comparifon with obfervations at a greater interval. The numerical method is therefore fcarce ever neceffary, but for comparing the Comet's computed Orbit with obfervation, when great accuracy is meant to be attained, by exact corrections of the errors found by diftant obfervations.

CHAPTER XIV.

Application of the Graphical Method to the Comet of 1769.
And first, on the Determination of the Curtate Distances
from the Earth, and the two Radii Vectores and Cord.

FIG. 14, 15. $\frac{241}{}$ § 1. WE now proceed to a detail of the whole method given in the foregoing Chapters, which we shall apply to the Comet of 1769. This Comet is chosen, as having had the Elements of its Orbit accurately determined by many celebrated Geometers, and also because it is used as an example, almost throughout Mr. Pingre's great work of the *Comêtographie.* From that work the three observations of the Comet now used, and the three others by which the Elements will be corrected, are extracted. The first are given in his Second Volume, Page 316; the latter in Page 368 of the same Volume.

As the Process is given in the utmost detail in the Example at the end of this part of the work; a few observations on the several steps of it, will render the whole perfectly clear.

§ 2. It may not be useless to premise, that for the Graphical part of the operation, a good Sector is absolutely requisite; which should be at least nine inches in length, and should have the

Line of Lines particularly, well divided. The Lines of Sines, &c. are not of so much consequence, as the value of any Line may be easily found in the Tables of Natural Sines, &c. It will however save some trouble, if they are well laid down on the Sector. As Lines greater than the Radius are often used; it will be extremely convenient to have the Line of Lines laid down on a long Sector, and continued to 15, instead of ending at 10, as is usually the case; as on a Sector so constructed, the value of Lines greater than Radius is found at once.

§ 3. As the deduction of the Longitude and Latitude of the Comet, from the observed Right Ascension and Declination, is an operation belonging to the general practice of Astronomy, it would be superfluous to mention it here. The First Articles in the Example therefore are: First, the Mean Times of the three Observations; with the three Places of the Sun, and its Distances from the Earth, for those times: which are found in the Ephemerides, or computed from the Solar Tables. Secondly, the Longitudes and Latitudes of the Comet, deduced from the Observations; then t, t', and t'', of which t, is the Time elapsed between the first and second Observation, reduced to Minutes; t', the Time between the second and third; and t'', the Time between the first and third. e, e', and e'', the three Elongations of the Comet from the Sun, or the difference of their Longitudes; and m, m', and m'', the Motion of the Comet in Longitude, between the first and second, second and third, and first and third Observations. Next to these, are the Numbers v', which, as the intervals of time between the Observations are equal, is the same as v, the versed Sine of half the Earth's Arch in her Orbit = half the difference of the first and third Longitude of the Sun;

X

FIG. 14, 15. L, L′, and L″, of which L, is the fum of the Logarithms of v', and the Sine of e'; L, the fum of the Logarithms of t'', the arithmetical complement of the Sine of m'', and the arithmetical complement of the Logarithm of t'; and L″, the fum of the Logarithm of t'', the arithmetical complement of the Sine of m'', and the arithmetical complement of the Logarithm of t. (In this example, as t, and t', are equal; of courfe L′ and L″, are alfo equal.) Laftly a; which is the fum of twice the Logarithm of t'', and the Conftant Logarithm: with the value of a, in numbers.

$\overset{247}{}$

§ 4. Having now prepared all the neceffary numbers, we proceed to the Graphical part of the operation. In the Example, (Figure 14), five inches are ufed as the Radius of the Earth's Orbit; which is drawn according to the directions given in Chapter 6, § 2, &c; and the three Places of the Earth T, T′, and T″, are fet off as there defcribed*.

$\overset{250}{}$

§ 5. The three Lines T E, T′ E′, and T″ E″, are drawn, making the Angles S T E, S T′ E′, and S T″ E″, equal to the elongations e, e', and e''; and prolonged at pleafure.

---

* It is not neceffary to take the trouble of drawing the Circles for the Ecliptic and Earth's Orbit, for if the three Lines S T, S T′, and S T″, be drawn, making the Angles T S T, and T′ S T, equal to the differences between the firft and fecond, and fecond and third Longitudes of the Sun; and on thefe Lines the three refpective Diftances of the Earth from the Sun, be fet off from S, taking at pleafure any length for Radius, or the Earth's Mean Diftance; the points T, T″, and T″, will be more accurate than the other way; which, however, is moft intelligible, and may fometimes prevent miftakes.

§ 6. The Right Angle A T B, (Figure 15), being prepared
apart, the Lines F, F', and F", are drawn ; making the Angles
F T B, F' T B, and F" T B, equal to the three Latitudes of the
Comet, l, l', and l'', and prolonged at pleafure. All thefe
Angles are fet off from the Sector by the Line of Cords ; or
they may be found by the Line of Lines, and Tables of Natural
Sines, making a convenient length Radius, from 10 to 10 on that
Line, and finding the values of the Cords in parts of that Ra-
dius, from the Tables. It is well known that twice the Sine of
half an Arch, is the Cord of that Arch.

§ 7. Every thing being now prepared, we proceed to a firft
fuppofition for the Curtate Diftance of the Comet from the
Earth, at the Second Obfervation. For this purpofe having
marked off the Points T, t, T,", on the ftrait edge of a flip of
paper, as directed Chapter 8, § 5, we may confider that if
brought too near the Earth, the Cord intercepted between the
Lines T E, and T" E", will be much lefs than the Cord of the
Earth's Motion, though the Comet's mean diftance is rather
lefs than that of the Earth ; in which cafe its velocity is to that
of the Earth, as $\sqrt{2}$ to 1 ; or as 14 to 10, nearly. Carrying it
therefore further off the Earth, and keeping t, on the paper always a little nearer to the Sun than the Line T' E', with the
edge in fuch a pofition as to make T, and T", lie equally within
the Lines T E, and T" E", and the Cord of the Comet's Motion more than as 14 to 10, of the Earth's, as the Comet is evidently nearer the Sun, the further off it is placed from the Earth ;
and alfo making an allowance on the other hand for the real
Cord being greater than the projected one between the Lines
T E, and T" E"; combining all thefe confiderations, the Diftance

T′ P′, was chosen, equal to 0 . 280 parts of the Radius of the Earth's Orbit, which is taken as 1,000.

The Distance T′ P′, is next set off on the Line T B, (Fig. 15), from T, to P′; and from that point, a perpendicular is raised to the Line T F′, cutting it in C′. (The Distance C′ P′, being the Comet's perpendicular Distance from the Plane of the Ecliptic)*. Then taking S P′, from Figure 14, in the compasses, we set it off on Figure 15, from P′ to S′. The Length S′ C′, is the Radius Vector of the Comet, whose value is found on the scale in parts of the Earth's Orbit, and with its Logarithm, forms the first division of the Table of Operation in the Example.

§ 8. The Second Division is for the Length of P′ p, the Cord of the Arch of the Comet's Motion. The Rule for finding this, is given in Chapter 8, § 12, and the Example needs no explanation.

§ 9. Set off P′ p, from P′, to p, (Figure 14), in the direction P′ S ; and T′ t, equal to v', from T′, to t, on the Line T′ S.

* As the Lengths of T P, T′ P′, and T″ P″, with the other Lines deduced from them, differ very little in the three positions, in order to avoid confusion in the Figure, the Perpendiculars here mentioned are only drawn in Figure 15, for the third trial ; and the Lengths of the Lines in Figure 14 and 15, for each of the three positions, are distinguished by marks as explained in the Plate. It is also to be observed, that as the Points P, and P″, on the Line T B, Figure 15, are extremely near equality in all the trials, in order that they may be more distinct, the marks for P′, are made above the Line T B, and those for P″, below it.

The Diſtance t p, in parts of the Earth's Orbit; is the correſted FIG. 14, 15 Curtate Diſtance, to be uſed in the following operation, and is the next article in the Example.

§ 10. The next ſtep is to find the Angle y, by which m, and m', are to be correſted. The Rule by which y, is found, is given Chapter 8, § 9, 10. The operation is plain. From the Logarithm before prepared, L, the Logarithm of t p, is ſubtraſted, that is, the arithmetical complement of that Logarithm is added. Their ſum is ſought in the Sines. From this firſt Logarithm, is ſubtraſted the triple Logarithm of S′ C′, by adding its complement as before; the Sine of this ſum is alſo found in the Tables; and from it the firſt found Sine is ſubtraſted. The remainder is the correſtion y.

§ 11. This correſtion is now to be applied to the Numbers m, and m', according to the Rule given in Chapter 8, § 11. In the preſent caſe, S C′, is leſs than the Radius of the Earth's Orbit; the motion of the Comet is direſt; and in paſſing from P′ to p, the motion is according to the Order of the Signs; the correſtion is therefore to be added to m, and ſubtraſted from m', which is done in the Example under the title of correſting m, and m'.

§ 12. Having correſted m, and m', from them, T P, and T″ P″, are now to be found. The method of finding them is given, Chapter 8, § 16; and the proceſs in the example is obvious on inſpeſtion.

§ 13. From the T P, and T″ P″, thus found, the S C, and S C″, are to be obtained Graphically, in the ſame manner as

Y

FIG. 14, 15. SC', was before found from T'P'; viz. by setting off the Lengths T P, and T" P", on the Lines T E, and T" E", from T, and T", to P, and P", (Figure 14.), and from T, to P, and P", (Figure 15); and raising the Perpendiculars P C, and P" C", (Figure 15), cutting the Lines F, and F". Then taking S P, and S P", (Figure 14), in the compasses, and setting them off on the Line T B, (Figure 15), from P, and P", as to S, and S"; the respective Lengths S C, and S" C", are the Radii Vectores required, whose value is found on the scale, and are the two first numbers in the division of the process bearing their name.

§ $\overline{14}^{260}$. The Line C C", which is the Cord of the Comet's Motion, is found as directed, Chapter 8, § 18, by setting off on the Line P" C", (Figure 15), (the greater Distance of the Comet from the Plane of the Ecliptic), the Line P C, (its lesser distance from that Plane), from C", towards P", as to I. (The Line I P", is the difference of the two distances from the Plane of the Ecliptic. Had the Latitudes been of different denominations, the sum of the distances must have been taken). Then from Figure 14 take P P", and set it off on Figure 15, from P", to E, then the Line I E, will be the Cord of the Comet's Motion, C C", (Figure 1). This quantity follows the two S C, and S C", in the Example.

§ $\overline{15}^{260}$. Having now obtained the two Radii Vectores, and the Cord of the Comet's Motion, as deduced from the first assumed T' P'; it remains to compare them with the formula a. The process for this forms the Third Great Division of the Example, subdivided into its several steps. A bare inspection of these, adverting at the same time to the Rules given in Chapter 8, § 18, must suffice for understanding them. The sum of S C,

and S C″, is b; whose Logarithm added to the doubled Logarithm of C C″, (which is c), gives the first part of the equation
$b c^2$: whose number stands opposite to it. The Second part is
$\frac{c^4}{12\,b}$. The sum of the Logarithms of b, and 12, (which last is

constant), gives 12 b; and this sum subtracted from the Qua-

drupled Logarithm of c, gives the Logarithm of $\frac{c^4}{12\,b}$. Its num-

ber is opposite. From $b c^2$ in numbers, $\frac{c^4}{12\,b}$ also in numbers, is

in the last division subtracted. The remainder is obviously

$b c^2 - \frac{c^4}{12\,b}$, which, if the position had been rightly taken, would

have been equal to the Number a; but it is not so, being con-
siderably less. Subtracting it therefore from a, we have the error of the first Position, which is called g, and has a negative
mark prefixed, the equation being less than a.

$\S\ \overset{261}{16}.$ We now therefore proceed to a Second Position, by
assuming a Second Length for T′ P′. In order to do this properly, we consider the drawing, and shall soon see that though
by taking T′ P′, greater than at first, the last Radius Vector
S C″, is shorter, yet as the first Radius Vector S C, is but little
altered, and that, rather in increase, the Cord C C″, is much
lengthened; and as a variation in the Cord has a much more

sensible effect in the Equation $b c^2 - \frac{c^4}{12\,b}$, than a variation in the

Radii Vectores; and as the value of the Equation was at first
too small, the T′ P′ ought now to be taken greater than at first.
The T′ P′ is therefore now assumed 0 . 300, and the whole operation is repeated. This being precisely the same as in the first
Position, can need no explanation. The error of the Equation

Fig. 14, 15. $bc^2 - \frac{c^4}{12b}$, is in this Pofition much diminifhed, but is ftill negative, and is called g'.

§ $\overset{263}{17}$. Having now the Errors g, and g', of the two Pofitions;
from them, by means of the Equation $\frac{g'\,b}{g-g'}$, (Chapter 8, § 20),
a third T′ P′, is obtained. The Procefs requires no explanation. To the correction of T′ P′, is alfo fubjoined a correction of S C, S C″, and C C″, by the fame formula; and a
comparifon of the Values fo corrected, with a; in order to fhew
how near thĕy are to truth, without further Pofitions. However,
in order to pufh the accuracy of this method as far as poffible,
a third Pofition is added, from the corrected T′ P′. This operation is the fame as the former, except in the part for the finding
y; in which the method of fubftituting S T′ p, for S T′ P′, given
in Chapter 8, § 10, is ufed. The other method is alfo given, in
order to fhew how little the Values of y, found by the two methods, differ. The error in this lâft Pofition is fo fmall, that a
Graphical procefs cannot be expected to approach nearer to the
truth; and the Radii Vectores, and Cord, are alfo extremely
near the truth; as may be feen by comparing them with the true
ones deduced from the accurate Elements determined by Profperin, (fee Pingrè, Vol. 2, Page 247), which are given at the
end of the Example. With thefe quantities we may therefore
proceed to the finding the Elements of the Orbit of the Comet.

CHAPTER XV.

Application of the Diſtances found, to the Determination of the Elements of the Comet of 1769, by Conſtruction.

§ 1. THE firſt Element to be ſought, is the Line of Nodes. FIG. 14, 15. For this purpoſe having the laſt determined diſtances, P C, = 0.114; P″ C″, = 0.135; and P P″, 0.201; we ſet off P C, (Figure 15), on P″ C″, from C″, to I; and P P″, from P″, to E; and draw I E. Parallel to the Line I E, from C″, we draw the Line C″ R; meeting the Line T B, in R. Having then prolonged the Line P P″, (Figure 14), at pleaſure; from P″, ſet off P″ R, (from Figure 15), whoſe value is 1.260°, to R. From R, to S, draw the Line R S. This is the Line of Nodes of the Comet's Orbit; and the point N, where it cuts the Ecliptic, gives the Longitude of the Node $11 \overset{s}{.} 2\overset{o}{5} . \overset{\prime}{0}$ As the Latitudes in our Example are South increaſing, it is evident that this is the Deſcending Node; the place of the Aſcending Node, is generally given in the Tables of the Elements of Comets, and it is Six Signs from the Deſcending. Its Longitude is therefore . $\overset{s}{5} . \overset{o}{2}\overset{}{5} . \overset{\prime}{0}$

§ 2. In caſes like the preſent, where the difference of the Lines P C, and P″ C″, is but ſmall, and of courſe the Lines

Z

FIG. 14, 15. C″ R, and P″ R, are very long; it is better to find P″ R, numerically, by the following proportion: P″ C″ — P C = P″ I : P P″ :: P″ C″ : P″ R. In the prefent Example, this proportion is, 0.21 : 0.201 :: 0.135 : 1.292. This differs only 0.032 from the P″ R, found above by conftruction only; and by applying this P″ R, as directed in the laft §, to the prolonged P P″, we get the Line of Nodes in $\overset{s}{5}$. $\overset{\circ}{24}$. $\overset{\prime}{45}$, differing only $15'$ from the former determination.

§ $\overset{269}{3}$. The next Element to be determined, is the Inclination of the Orbit. To find this, we let fall the Perpendiculars P D, and P″ D″, (Figure 14), from P, and P″, on the Line of Nodes, meeting it in D, and D″. The Lengths of thefe Lines are 0.120, and 0.143. Setting off thefe Lines on Figure 15, from P, and P″, refpectively, to D, and D″; we draw the Lines D C, and D″ C″. The Angles P D C, or P″ D″ C″, are the Inclination of the Orbit. The firft of thefe in the Example, is $\overset{\circ}{43}$. $\overset{\prime}{15}$. the latter is $\overset{\circ}{43}$ *.

§ 4. Making ufe of the Afcending Node found by the Numerical determination of P″ R, in § 2, we obtain P D, 0.125; and P′ D″, 0.148; which applied to the Sector in the man-

---

* The inclination may alfo be found, by making P D, or P″ D″, Radius on the Line of Tangents on the Sector; then P C, or P″ C″, will give the inclination. This method has this advantage; that a large Scale may be ufed, as the value of the Lines P C, and P D, has been found before in parts of the Radius of the Earth's Orbit; and of courfe greater accuracy obtained when, as in the prefent cafe, thefe Lines are fmall. In this way P D C, and P″ D″ C″, both gave $\overset{\circ}{43}$. $\overset{\prime}{30}$ for the inclination.

ner given in the Note; give for the Inclination; P D C, 42 . 20
and P″ D″ C″, 42 . 25 .

§ 5. In order to find the Perihelion diſtance, and its Longitude; it is firſt neceſſary to lay down the Comet's true places in its Orbit. This is done by ſetting off on D P, and D″ P″, prolonged, (Figure 14), the Lines D C, and D″ C″, of Figure 15. Then C, and C″, (Figure 14), will be the true places of the Comet in its Orbit, which is here ſuppoſed to be as it were applied or laid down on the Plane of the Earth's Orbit. If the operation for the Line of Nodes and Inclination has been rightly done, the Lines S C, S C″, and C C″, thus obtained; will be the ſame as were found before in the laſt Poſition.

§ 6. As in this Example it is evident that the method given by Boſcovich, (Chapter 9, § 8), for finding the Perihelion diſtance is totally inapplicable; the method deſcribed in the Note on that §, is made uſe of. The Semicircle C L C″, is deſcribed on C C″, (Figure 14), and the difference of the two Radii Vectores, is ſet off from C″, to L; and the Line C″ L, is drawn. Parallel to this the Line, S K, is drawn, paſſing through the Sun, and cutting the Ecliptic in K, with a Longitude of 10 . 23 . 55. This is the Axis of the Orbit, and Six Signs from K, is the Longitude of the Perihelion on the Comet's Orbit; that is, 4 . 23 . 55.

§ 7. To find the Perihelion diſtance, the Line C L H, is drawn, falling on the Axis at H, at Right Angles to it. From H, on the Line K S, produced beyond S, to X, the Radius

FIG. 14, 15. Vector S C, is set off to X. Then X, is double the Perihelion distance of the Comet, and S X, bissected in V, gives S V, the Perihelion distance; which in the Example is 0 . 1205. As it is always better to apply a short numerical procefs, where the distances are small, as in this case, S V, is; the Length of H S, is found on the Scale, which is here, 0 . 783. This subtracted from the Radius Vector, S C, = 1 . 024, leaves 0 . 241. = S X. Half of this is the Perihelion distance = 0 . 1205.

$\underset{272}{\quad}$

§ 8. For the time of the Comet's arrival at the Perihelion, the method given in Chapter 6, § 13, is ufed. The Line B A B, is drawn at Right Angles to the Axis; and bissecting the Perihelion distance, S V, in A. Then having bissected the Radii Vectores, S C, and S C''; in Y, and Y''; we draw the Lines Y O, and Y'' O'', perpendicular to S C, and S' C''; and meeting the Line B A B, in O, and O''. Then O O'', is the Comet's motion, for eight days, whose value on the Scale is 0 . 209. The value of A O is 0 . 863. Then say

209 : 8 Days : : 863 : 33 . 0335 Days. Which being added to the Epoch of O,

	D. H.		
September 4.	14	—	. 4 . 5833
Days		—	33 . 0335
Gives		—	37 . 6168

		D. H. M.
Equal to October	—	7 . 14 . 48 for the Arrival at the Perihelion.

§ 9. The time of the Arrival at the Perihelion may alfo be found, by the method given in Chapter 6, § 11, as follows.

Perihelion Diſtance   o . 1205   Log.   9. 0809870   FIG. 14, 15.

Its half   9. 5404935

Log. of the time the Comet, whoſe Perihelion Diſtance is = the Earth's, takes
to move 90° of Anomaly—Conſtant   2. 0398717

Time the Comet takes to move 90° Anomaly o. 6613522

Whoſe Number is   4. 58514 Days.

Then ſay

DAYS.   DAYS.

S V = o. 1205 : 4. 58514 : : A O = o. 8630 : 32 . 83797

which being added to the Epoch of O,

	D. H.		
September	4 . 14	=	4 . 58330
Days	—	—	32 . 83797
	Gives	—	37 . 42130

D. H. M. S.

Equal to October   —   —   7 . 10 . 6 . 42 for
the Arrival at the Perihelion.

§ 10. The Sixth and laſt Element of the Orbit of the Comet,
is the direction of its motion ; which a bare inſpection ſhews to
be direct. The Table at the end of this part of the Example,
collects all theſe Elements together, as they have been determined by each operation explained above.

A a

CHAPTER XVI.

Determination of the Place of the Comet, for another given Time.

§ 1. **I**N order to obtain the place of the Comet for any other time, the Orbits of the Earth and Comet muſt be drawn in their true Proportions from the determined Elements, as directed in Chapter 6, § 1, 2, &c. In the preſent Example we have however made uſe of the Elements found trigonometrically, as being more exact than thoſe already obtained by conſtruction.

They are as follow :

	s	°		
Place of the Aſcending Node —	5 .	25 .	9 .	33
Inclination of the Orbit —		40 .	59 .	50
Place of the Perihelion on the Orbit	4 .	24 .	37 .	51
Perihelion Diſtance — —	0 .	1203815		
Its Logarithm — —	9 .	0805598		

	D.	M.	S.
Time of the Arrival at the Peri-⎱ October	7 .	12 . 11 . 9	
helion — — ⎰ or	7 .	5077	
Motion.	Direct.		

§ 2. Figure 16, repreſents the Cometary Orbit, both in its real dimenſions, and as applied to the Plane of the Ecliptic, and divided to every fifth day, according to the method given in Chapter 6, § 9. For this purpose, the time the Comet takes to move 90° of Anomaly, is thus found.

Logarithm of Perihelion Distance	—	9 . 0805598
Its half	— —	9 . 5402799
Constant Logarithm	—	2 . 0398720
Sum	—	0 . 6607117

Whose Number — — 4 . 5784, is the Number of Days and Decimals employed by the Comet to move 90° of Anomaly, and by the centre O, to move A'O, = the Perihelion Distance, on the Line B B', (Fig. 8). Taking therefore the Perihelion Distance in the compasses, we apply it to 458 on the Line of Lines, on the Sector; then the even Numbers, 1, 2, 3, &c. are the daily Motions of the Centre O, on the Line B B'; whence Arches struck with the Radii O S, will cut the Orbit in the daily corresponding Places of the Comet. The Lines perpendicular to the Line of Nodes, termi- nate in points; which are the Comet's Places projected on the Plane of the Ecliptic: and the other Points, on the same Lines, are the perpendicular Distances of the Comet from the Plane of the Ecliptic. These Points are laid down in the manner de- scribed in the Note to Chapter 10, § 3. The former being the Cosines of the Inclination of the Orbit, taking the perpendicu- lar Distance of the Comet from the Line of Nodes, as Radius; and the latter, the Sines of the Inclination, to the same Radius.

277
§ 3. As much of the Ecliptic, and Orbit of the Earth are re- presented, as the Earth passed through during the time of the appearance of the Comet. To have drawn the whole would have needlessly enlarged the Plate.

Fig. 16.

§ 4. The Orbits of the Earth and Comet being thus prepared, let the Place of the Comet be required for the Third of
November, at 6 . 30, or 3,2708 Mean Time. The Place of
the Earth as given by the Ephemeris, is 1 . 11 . 38 . 54;
which is set off on the Ecliptic, at Q, and the corresponding
Point in the Orbit of the Earth, at T. The Distance of Time,
from the Time of the Comet's arrival at the Perihelion, October
7 . 12 . 11 . 9, or 7,5077; is 26,7631. Having found the
Point O, in the Line B B', answering to that Distance, the
Distance O S, set off from O, to C, gives the Place of the Comet
in its Orbit. The Line C D, let fall from C, perpendicular to
the Line of Nodes, is taken for Radius; to which the Cosine
of the Inclination, set off from D, towards C, gives P, the Place
of the Comet reduced to the Ecliptic; and the Sine of the Inclination to the same Radius, set off likewise from D, gives
D ☊, the perpendicular Distance of the Comet from the Plane
of the Ecliptic.

§ 5. The Place of the Comet in its Orbit being thus found
for the required Time; from the Point T, to P, draw the
Line T P; and a Parallel to it, passing through the Sun, will
cut the Ecliptic in the Geocentric Longitude required. Or by
the Line of Cords on the Sector, find the value of the Angle
S T P, which is the Comet's Elongation from the Sun; equal
in this Example to 25 . 46, E; which being added to the
Longitude of the Sun 7 . 11 . 38 . 54, gives the Longitude
of the Comet 8 . 7 . 25.

278

§ 6. For the Geocentric Latitude, take T P, and make it
Radius on the Line of Tangents. Then with D ♌, in the compasses, find the Angle it subtends, which here is, $21 . 30'$. This
is the Geocentric Latitude of the Comet *.

§ 7. The Longitude and Latitude of the Comet, by observation, were; Longitude, $8^s . 6^\circ . 39' . 32''$, and Latitude,
$21^\circ . 0' . 34''$. The errors therefore of the Place of the Comet,
deduced from the Elements obtained from the observations of
September, are, $+ 0^\circ . 45\frac{1}{2}'$ in Longitude; and $+ 0^\circ . 29\frac{1}{2}'$ in
Latitude. These errors, though confiderable, are not greater
than might be expected in Elements deduced from obfervations
of a Comet, whofe Perihelion Diftance is fmall; when compared
with an obfervation made at an interval of two months, and in
the other branch of the Orbit. The Second Part of this Work
will fhew how to correct thefe approximate Elements, and determine the Orbit of the Comet to the utmoft precifion of which
the obfervations are fufceptible.

* By this method were all the Places of the Comet, whofe return was
expected in the year 1789, found; on the large Plate of the Comet's
Orbit, annexed to the Tables of its apparent Places, publifhed in the
year 1788.

B b

CHAPTER XVII.

Application of the Trigonometrical Method to the Comet of 1769. And first, for the Determination of the two Curtate Distances from the Earth, and the Radii Vectores and Cord.

§ 1. ³⁰⁰ THE Elements of the Comet obtained in the foregoing Chapters, are as near the truth as can be expected from a Graphical operation. If therefore a greater degree of accuracy is required, recourfe muft be had to the longer but more exact method of Trigonometrical calculation; given in Chapter 11. This procefs is given at large in the Example; and but few obfervations will be requifite on the feveral fteps of it.

§ 2. ³⁰¹ The quantities previoufly to be prepared are the fame as thofe ufed in the Graphical procefs, and given in Part the Firft of the Example.

§ 3. ³⁰² The firft thing therefore now to be done, is to affume the T' P', given by the laft Graphical procefs, which in the prefent Example is 0. 305, and from thence to determine the correction of the Second Longitude, by the method given in

Chapter 4, § 8. The firſt thing to be obtained is the Angle T′ S P′, (Figure 4 and 14). For this we have T′ P′, aſſumed, T′ S, being the Earth's diſtance from the Sun, and the contained Angle S T′ P′, the Second Elongation of the Comet. The Angle T′ S P′, and the Side S P′, are found by the Theorems given in Chapter 11, § 4.

Having the Angle T′ S P′; we thence find the Side S P′.

§ 4.
302
  We next, in the Triangle T′ P′ C′, right-angled at P′, find the Side P′ C′, from the Side T′ P′, and the Angle P′ T′ C′, (= l', the Geocentric Latitude), given: by the Rule of Chap. 4, § 8, or Chap. 11, § 3.

§ 5.
304
  Having now in the Triangle S P′ C′, right-angled at P′, the two Sides S P′ and P′ C′; the third Side S C′, is to be found by the two Theorems given in Chapter 11, § 7.

Having now obtained S C′, and S P′, accurately, the verſed Sine of the Cometary Cord, P′ p, is found as in the Graphical proceſs.

§ 5.
305
  Having now P′ p; S p, is found by ſubtracting P′ p, from S P′; and S t, is in like manner found by ſubtracting v', found before in the Graphical proceſs, from S T′.

§ 6.
306
  Having now in the Triangle S t p, the Sides S t, and S p; and the included Angle T′ S P′; the Angle S t p, is found by the proportion uſed before for the Angle T′ S P′. The difference of S t p, from the Second Elongation, e', is the cor-

rection, y, which is to be applied to m, and m', as directed before in the Graphical procefs.

308

§ 7. Having now the corrected m, and m; the length of the Side t.p, is next to be obtained. For this, in the Triangle S t p, we have the Angle S t p, its oppofite Side S p, and the Angle t S p = T′S P′. Whence the Side t p, is found, as S P, has been already found.

309

§ 8. Having t p, and m, and m'; T P, and T″ P″, are deduced from them by the quantities L′ and L″, as in the Graphical procefs.

309

§ 9. Having T P, and T″ P″, and l, and l''; P C, and P″ C″, are refpectively found from them; as P′ C′, has been already found.

309

§ 10. The Angles T S P, and T″ S P″, are alfo found, as T′ S P′ has been before.

310

§ 11. The Sides S P, and S P″; and S C, and S C″; are alfo found by the fame procefs which before gave the Sides S P′, and S C′.

312

§ 12. The length of P P″, is now to be found, in order to obtain which, the Angle P S P″, is neceffary. Having the Angles at the Sun, T S T″, T S P, and T″ S P″, the Angle P S P″, is eafily got. In the prefent cafe it is = T S T″ + T″ S P″ — T S P; in other cafes it may vary, but the infpection of the Figure will be a fure Guide. Having now in

the Triangle P S P", the Sides S P, and S P", and the included Fig. 4, 14, 15.
Angle P S P", the Angle S P P", or S P" P, muſt be found,
as the Angles T S P, and T" S P", were found before. (We
have here found S P P"). Then P P", will be obtained by the
proportion which gave S P, and S P'.

§ $\overset{312}{13}$. From P P"; C C", will now be found by means of the
two perpendicular diſtances of the Comet, from the Plane of the
Ecliptic, P C, and P" C". As the Latitudes are of the ſame
denomination, their difference muſt be taken; had the Latitudes been of oppoſite denominations, their ſum muſt have
been uſed. Then in the right-angled Triangle, whoſe Baſe
is P P", and Perpendicular, P" C" — P C = C" I; we find
the Hypothenuſe C C", by the Theorems given in Chapter
11, § 7, and 9.

§ $\overset{313}{14}$. Thus we at length have the Radii Vectores, S C, and
S C"; and the Cord of the Comet's motion, C C". The com-
pariſon with a in the Formula $b c^2 - \dfrac{c}{12\,b}$ is preciſely the ſame
as in the Graphical proceſs. The error of this poſition is +
0. 0010569. = g.

§ 15. To correct this error, another length of T' P' was aſſumed = 0. 303. This operation being in every reſpect ſimilar
to that in the firſt poſition, it is not given. The error of this
poſition is + 0. 0000549. = g'. The error of this poſition is
ſo ſmall, that we may ſafely uſe the Equation, $\dfrac{g'\,b}{g - g'}$, to obtain
the correct lengths of S C, S C", and C C". The operation is
preciſely the ſame as that in the Example to the Graphical operation, and needs no comment.

C c

§ 16. From the SC, SC'', and CC'', thus corrected, the Equation $bc^2 - \dfrac{c^4}{12\,b}$ is once more formed, and compared with a. The error is now reduced to $+ 0.00000178$. This may be confidered as abfolutely infenfible; and therefore the Numbers corrected by the Equation $\dfrac{g'\,b}{g-g'}$, may be applied to the computation of the Elements of the Orbit of the Comet.

§ 17. The laft divifion therefore of the Example contains the correction of all the quantities to be ufed in the computation of the Elements, by the Equation $\dfrac{g'\,b}{g - g'}$.

§ 18. In order to try how far the correction of the Numbers is to be depended on, the Angle SPP'', and the Side PP'', were computed, from the corrected SP, and SP''. The Angle SPP'', was found by this trial $14.\overset{\circ}{}\,23.\overset{'}{}\,57\overset{''}{}$, differing from the Angle corrected by the Equation, only $2''$, a quantity quite infenfible; and the length of PP'', 0.200820; differing from the corrected PP'', only 0.000002. Which may be confidered as an abfolute agreement.

CHAPTER XVIII.

Determination of the Elements of the Orbit of the Comet of 1769, *by the Trigonometrical Method.*

§ 1. $\overset{333}{}$ **H**AVING corrected all the Numbers and Angles re- FIG. 14, 15. quifite for the computation of the Elements of the Orbit of the Comet; we fhall firft find the Longitude of the Afcending Node, according to the Rules given in Chapter 12, § 1. And firft we find P R. To P R, we add P P″. The fum is the Side P″ R. Having now in the Triangle P″ S R, the Angle S P″ P, and the Sides containing it, S P″, and P″ R; the Angle at the Sun, P″ S R, is found as directed in Chapter 12, § 1. Having the Angle P″ S R; in the prefent cafe, the Angle T″ S R, is found by fubtracting P″ S R, from T′ S P″. And in the prefent cafe, the fum of T″ S R, and the Earth's Longitude at the third Obfervation, T″, give the Longitude of the Node; which infpection of the Figure ufed in the former Graphical procefs, fhews to be the Defcending Node. The Afcending Node is Six Signs from it.

§ 2. $\overset{336}{}$ The Inclination is found by the method given in Chapter 12, § 2. The operation needs no comment.

§ 3. $\overset{337}{}$ The Anomaly of the Comet is next to be found at the time of the firft Obfervation, as it was approaching the Sun.

FIG. 14, 15. The Rule for finding it is given at length in Chapter 12, § 4. Had the Comet been receding from the Sun, the Anomaly at the third Obfervation would have been found, as the Rule univerfally gives the Anomaly of the greater Radius Vector.

§ 4. The Anomaly thus found, gives the two next Elements of the Orbit of the Comet, the Perihelion diftance, and time of arrival at the Perihelion. The Rules are given at length in Chapter 12, § 4, 5, and 8, and the procefs is fufficiently clear.

341

§ 5. As in the Example here given, the Perihelion diftance' is deduced from the Anomaly of the greater Radius Vector, and the Longitude of the Node from the leffer Radius Vector, it becomes neceffary to change the order of the Elements ; and to find the Time of the arrival at the Perihelion, previoufly to the place of the Perihelion. This, however, makes no difference whatever in the operation, and is only noticed in order to avoid the appearance of irregularity. The place of the Perihelion on the Orbit is therefore now to be found. The procefs confifts of two parts: Firft, the finding the Angle $C''SR$, the Rule for which is given in Chapter 12, § 6. The Second part is for the Anomaly of the leffer Radius Vector, SC'', which is deduced from the Perihelion diftance, and time of arrival, by the Rule given in the directions for the ufe of the Tables. This Angle, which is $C''SV$, being in the prefent cafe, added to $C''SR$, gives RSV, the Angle between the Axis of the Orbit, and the part of the Line of Nodes adjacent to $C''SR$. As the infpection of the Figure (Figure 14), fhews that the Longitude of the Afcending Node is forwarder in refpect of the Signs, than the Longitude of the Axis; the place of the

Perihelion would have been found equally, by fubtracting FIG. 14, 15. R S V, from 180°; which would have given the Angle between V S, and the Line of Nodes, towards the Afcending Node; this Angle fubtracted from the Longitude of the Afcending Node, would have given the Longitude of the Perihelion on the Orbit. It is evident, that in many cafes the Angles will be differently placed with refpect to each other; but the Figure conftructed for the Graphical procefs, will be ever a fure guide in the Trigonometrical operation, and fhew when to add, and when to fubtract the Angles.

§ 6. The Sixth Element needs no comment. The Elements obtained by this method are as in the Graphical procefs, collected into one Table at the end of the Example.

D d

EXAMPLE of the GRAPHICAL OPERATION for the ORBIT of the COMET of 1769.

PART THE FIRST. *Obfervations ufed, and Quantities prepared.*

Times of Obfervation. SEPTEMBER.	Longitudes of Comet.	Latitudes of Comet. SOUTH.	Longitudes of Sun.
D. H. M. S.	S. ° ′ ″	° ′ ″	S. ° ′ ″
4 . 14 . 0 . 0	2 . 20 . 56 . 11	l - 17 . 51 . 39	5 . 12 . 42 . 5
8 . 14 . 0 . 0	3 . 11 . 0 . 54	l' - 22 . 5 . 2	5 . 16 . 35 . 31
12 . 14 . 0 . 0	4 . 4 . 19 . 22	l'' - 23 . 43 . 55	5 . 20 . 29 . 20

Diftances of Earth.	$t, t',$ and t''.	Elongations. WEST.	$m, m',$ and m''.
	M.	° ′ ″	
S T 1 . 00740	t - 5760	e - 81 . 45 . 54	m 20 . 4 . 43
S T' 1 . 00615	t' - 5760	e' - 65 . 34 . 37	m' 23 . 18 . 28
S T'' 1 . 00500	t'' - 11520	e'' - 46 . 9 . 58	m'' 43 . 23 . 11

QUANTITIES PREPARED.

v	L	L	a
° ′ ″			
Arch 7 . 47 . 15	v 7 . 362967	t'' - 4 . 061452	t'' - 4 . 0614525
Half 3 . 53 . 37	Sin. e' 9 . 959290	Sin. m'' a.c. 0 . 163111	t'' doubled 8 . 0228050
v 7 . 362967	L 7 . 322257	t' ac. - 6 . 239578	Conftant 0 . 7564990
v' is the fame.		L' - 0 . 464141	a - 8 . 8794040
		L'', is the fame.	in numbers
			a - 0 . 075754

PART THE SECOND. *Trials for* S C, S C″, *and* C C″, *and the Comparison of the Equation* $b c^2 - \dfrac{c^4}{12 b}$, *with* a.

	FIRST POSITION.		SECOND POSITION.	
For S C′.	LOGARITHMS.	NUMBERS.	LOGARITHMS.	NUMBERS.
T′ P′ - -	- -	0 . 280	- -	0 . 300
P′ C′ - -	- -	0 . 114	- -	0 . 122
S P′ - -	9 . 967548	0 . 928	9 . 9652017	0 . 923
S C′ - -	9 . 970812	0 . 935	9 . 9680157	0 . 929
S C′ tripled	9 . 912436		9 . 9040471	
For P′ p.				
v - - -	7 . 362967		7 . 3629670	
S P′ - -	9 . 967548		9 . 9652017	
S C′ tripled, a. c.	0 . 087564		0 . 0959529	
P′ p - -	7 . 418079	0 . 00268	7 . 4241216	0 . 002656
t p - - -	9 . 444044	0 . 278	9 . 4742163	0 . 298
For y.				
t p, a. c. - -	0 . 555956		0 . 5257837	
L - - -	7 . 322257		7 . 3222570	
Sin. - -	7 . 878213	0 . 25 . 58	7 . 8480407	0 . 24 . 14
S C′, tripled, a. c.	0 . 087564		0 . 0959529	
Sin. - -	7 . 965777	0 . 31 . 46	7 . 9439936	0 . 30 . 14
y - - -	- -	0 . 5 . 48	- -	0 . 6 . 0
For correcting *m*, and *m*′.				
m - - -	- -	20 . 4 . 43	- -	20 . 4 . 43
y + - -	- -	0 . 5 . 48	- -	0 . 6 . 0
m - - -	- -	20 . 10 . 31	- -	20 . 10 . 43
m′ - -	- -	23 . 18 . 28	- -	23 . 18 . 28
y — - -	- -	0 . 5 . 48	- -	0 . 6 . 0
m′ - -	- -	23 . 12 . 40	- -	23 . 12 . 28

For T P.	LOGARITHMS.	NUMBERS.	LOGARITHMS.	NUMBERS.
	FIRST POSITION.		SECOND POSITION.	
t p - - -	9 . 444044		9 . 4742163	
Sin. m' - -	9 . 595627		9 . 5955700	
L' - - -	0 . 464141		0 . 4641410	
T P - -	9 . 503812	0 . 3190	9 . 5339273	0 . 3419
S P - -	- -	1 . 0080	- -	1 . 0180
P C - -	- -	0 . 1060	- -	0 . 1120
For T″ P″.				
t p - - -	9 . 444044		9 . 4742163	
Sin. m - -	9 . 537687		9 . 5377550	
L″ - - -	0 . 464141		0 . 4641410	
T″ P″ - -	9 . 445872	0 . 2791	9 . 4761123	0 . 2993
S P″ - -	- -	0 . 8360	- -	0 . 8270
P″ C″ - -	- -	0 . 1240	- -	0 . 1340
P P″ - -	- -	0 . 1840	- -	0 . 1970
S C - -	- -	1 . 0220	- -	1 . 0240
S C″ - -	- -	0 . 8430	- -	0 . 8370
C C″ - -	- -	0 . 1860	- -	0 . 1980

E e

Formation of the Equation b c² − $\frac{c^4}{12b}$, and its Comparison with a.

For bc^2.	FIRST POSITION. LOGARITHMS.	NUMBERS.	SECOND POSITION. LOGARITHMS.	NUMBERS.
$e = CC''$	9.2695129	0.1860	9.2966652	0.1980
$b = SC + SC''$	0.2706788	1.8650	0.2697464	1.8610
c^2 - - -	8.5390258		8.5933304	
bc^2 - -	8.8097046	0.0645215	8.8630768	0.072959
For 12 b.				
b - - -	0.2706788		0.2697464	
12 Conſtant -	1.0791810		1.0791810	
$12 b$ - -	1.3498598		1.3489274	
For $\frac{c^4}{12b}$.				
c^4 - - -	7.0780516		7.1866608	
$12 b$ - - -	1.3498598		1.3489274	
$\frac{c^4}{12 b}$ - - -	5.7281918	0.0000535	5.8377334	0.0000688
For $bc^2 - \frac{c^4}{12b}$, and comparing it with a.				
bc^2 - - -	- -	0.0645215	- -	0.0729590
$\frac{c^4}{12 b}$ - -	- -	0.0000535	- -	0.0000688
$bc^2 - \frac{c^4}{12 b}$ -	- -	0.0644680	- -	0.0728902
a - - -	- -	0.0757540	- -	0.0757540
g — - -	- -	0.0112860	g' — -	0.0028638

PART THE THIRD. *Correction of* T'P', SC, SC", *and* CC", *of the Second Position, by the Formula* $\dfrac{g'h}{g-g'}$. *And Comparison of the Corrected Quantities with* a.

	LOGARITHMS.	NUMBERS.		NUMBERS.
g — - -	- -	0 . 011286		
g' — - -	7 . 4568213	0 . 002863		
$g - g'$ —. a. c. -	2 . 0745332	0 . 008423		
$\dfrac{g'}{g-g'}$ + - -	9 . 5313545			
$h\,T'P'$ + -	8 . 3010300	0 . 020000	T'P' -	0 . 3000
$\dfrac{g'-h\,T'P'}{g-g'}$ + -	7 . 8323845	0 . 006798	Corr. +	0 . 006798
			T'P' -	0 . 306798
$\dfrac{g'}{g-g'}$ - -	9 . 5313545			
$h\,SC$ + - -	7 . 3010300	0 . 0020000	SC -	1 . 024
$\dfrac{g'h\,SC}{g-g'}$ + -	6 . 8323845	0 . 0006798	Corr. +	0 . 0006798
			SC -	1 . 0246798
$\dfrac{g'}{g-g'}$ - - -	9 . 5313545			
$h\,SC''$ — - -	7 . 7781513	0 . 00600	SC" -	0 . 837
$\dfrac{g'h\,SC''}{g-g'}$ — -	7 . 3095058	0 . 00204	Corr —	0 . 00204
			SC" -	0 . 83496
$\dfrac{g'}{g-g'}$ - -	9 . 5313545			
$h\,CC''$ + - -	8 . 0791812	0 . 012000	CC" -	0 . 198
$\dfrac{g'h\,CC''}{g-g'}$ + -	7 . 6105357	0 . 004079	Corr. +	0 . 004079
			CC" -	0 . 202079

Formation of the Equation $bc^2 - \dfrac{c^4}{12b}$ *with the Quantities corrected, and its Comparison with* a.

For bc^2.	LOGARITHMS.	NUMBERS.
$c = CC''$ -	9 . 3055211	0 . 202079
$b = SC + SC''$	0 . 2694288	1 . 85964
c^2 - -	8 . 6110422	
bc^2 - -	8 . 8804710	0 . 075940
For $12\,b$.		
b - - -	0 . 2694288	
12 Constant -	1 . 0791812	
$12\,b$ - -	1 . 3486100	
For $\dfrac{c^4}{12b}$.		
c^4 - - -	7 . 2220844	
$12\,b$ - -	1 . 3486100	
$\dfrac{c^4}{12\,b}$ - -	5 . 8734744	0 . 00007473
For $bc^2 - \dfrac{c^4}{12b}$, and comparing it with a.		
bc^2 - -	- -	0 . 0759400
$\dfrac{c^4}{12b}$ - -	- -	0 . 0000747
$bc^2 - \dfrac{c^4}{12b}$ -	- -	0 . 0758653
a - - -	- -	0 . 0757540
Error + - -	- -	0 . 0001113

The Length of T′ P′ being taken, o . 3068 ; a Third Trial was made. The Proceſs in this Poſition being ſimilar to that in the two former Poſitions, except in the method of obtaining *y*, no part of the Operation is given but that ; and *y* is alſo found by the method uſed in the former Poſitions.

For *y*.			LOGARITHMS.	NUMBERS.
S T′ p Sin.	-	-	9 . 9557890	64 . 35 . o
v	-	-	7 . 3629670	
t p	-	a. c.	o . 5161275	
Sin.	-	-	7 . 8348835	o . 23 . 30
S C′ tripled	-	a. c.	o . 0931509	
Sin.	-	-	7 . 9280344	o . 29 . 8
y	-	-	- -	o . 5 . 38
For *y* as before.				
t p	-	a. c.	o . 5161275	
L	-	-	7 . 3222570	
Sin.	-	-	7 . 8383845	o . 23 . 42
S C′ tripled	-	a. c.	o . 0931509	
Sin.	-	-	7 . 9315354	o . 29 . 22
y	-	-	- -	5 . 40

In this Poſition the Quantities obtained were

S C	-	-	-	I . 0235
S C″	-	-	-	o . 8350
C C″	-	-	-	o . 2020

And with theſe Quantities the Equation $b c^2 - \dfrac{c^4}{12 b}$, being formed, the reſult was

$b c^2 - \dfrac{c^4}{12 b}$	-	-		o . 0757586
a	-	-	-	o . 0757540
Error	-	-	-	$+$ o . 0000046

F f

The Lengths of S C, S C″, and C C″, as computed from the Corrected Elements by Profperin, given in the *Comêtographie* of Pingrè, Vol. 2, Page 247; are

S C	-	-	-	1 . 02336
S C″	-	-	-	0 . 83526
C C″	-	-	-	0 . 20195

The differences from thofe found above are

S C	-	-	-	— 0 . 00014
S C″	-	-	-	+ 0 . 00026
C C″	-	-	-	— 0 . 00005

Elements of the Orbit of the Comet of 1769, as determined by the Graphical Method.

				S. ° ′
1.	Place of Afcending Node by § 1	-		5 . 25 . 0
	Again by § 2	-	-	5 . 24 . 45
2.	Inclination of the Orbit by § 3	-		43 . 0
	Again	-	-	43 . 15
	Again by Note on § 3	-	-	43 . 30
	Again by § 4	-	-	42 . 20
	Again	-	-	42 . 25
3.	Longitude of the Perihelion	-		4 . 23 . 55
	Again by a larger Drawing	-		4 . 24 . 23
4.	Perihelion Diftance	-	-	0 . 1205
	Again by a larger Drawing	-		0 . 1180

				D. H. M.
5.	Arrival at the Perihelion	-	October	7 . 14 . 48
	Again by § 9	-	-	7 . 10 . 7
	Again by a larger Drawing	-	-	7 . 18 . 58
6.	Motion of the Comet.	-	-	Direct.

EXAMPLE of the TRIGONOMETRICAL OPERATION for the ORBIT of the COMET of 1769.

The Obſervations, and Quantities prepared, the ſame as in the Graphical Proceſs.

PART THE FIRST.

Trial for S C, S C″, *and* C C″; *and the Compariſon of the Equation*

$$b c^2 - \frac{c^4}{12\,b} \quad \text{with a.}$$

For T′ S P′.			LOGARITHMS.	NUMBERS.
S T′ P′ = e′	-	-	- -	65 . 34 . 37
Two remaining Angles	-		- -	114 . 25 . 23
Half of them	-	-	- -	57 . 12 . 41
T′ P′	-	-	- -	0 . 30500
S T′	-	-	- -	1 . 00615
S T′ + T′ P′	-	a. c.	9 . 8823477	1 . 31115
S T′ — T′ P′	-	-	9 . 8458109	0 . 70115
Tangent	-	-	10 . 1909917	57 . 12 . 41
Tangent	-	-	9 . 9191503	39 . 41 . 50
T′ S P′	-	-	- -	17 . 30 . 51

For S P′.				
Sin. T′ S P′	-	a. c.	0 . 5215200	
T′ P′	-	-	9 . 4842998	
Sin. e′	-	-	9 . 9592885	
S P′	-	-	9 . 9651083	0 . 92281

For P′ C′.				
T′ P′	-	-	9 . 4842998	
Tangent l′	-	-	9 . 6082374	22 . 5 . 2
P′ C′	-	-	9 . 0925372	0 . 12378

For S C′.			LOGARITHMS.	NUMBERS.
S P′	-	a. c.	0 . 0348917	
P′ C′	-	-	9 . 0925372	
Tangent P′ S C′	-	-	0 . 1274289	7 . 38 . 18
Sin. P′ S C′	-	a. c.	0 . 8764300	7 . 38 . 18
P′ C′	-	-	9 . 0925372	
S C′	-	-	9 . 9689672	0 . 93104
For P′ p.				
S C′ tripled	-	-	9 . 9069016	
v′	-	-	7 . 3629670	
S P′	-	-	9 . 9651083	
S C′ tripled	-	a. c.	0 . 0930984	
P′ p	-	-	7 . 4211737	0 . 002637
For S′ p.				
S P′	-	-	- -	0 . 92281
P′ p	-	-	- -	0 . 00264
S p	-	-	- -	0 . 92017
For S t.				
S T′	-	-	- -	1 . 00615
v′	-	-	- -	0 . 00231
S t	-	-	- -	1 . 00384
For S t p.				
T′ S P′	-	-	- -	17 . 30 . 51
Two remaining Angles	-	-	- -	162 . 29 . 9
Half of them	-	-	- -	81 . 14 . 35½
S p	-	-	- -	0 . 92017
S t	-	-	- -	1 . 00384
S t + S p	-	a. c.	9 . 7157926	1 . 92401
S t — S p	-	-	8 . 9225698	0 . 08367
Tangent of Half Sum	-	10 . 8123704	81 . 14 . 35
Tangent	-	-	9 . 4507328	15 . 45 . 54
S t p	-	-	- -	65 . 28 . 41

	LOGARITHMS.	NUMBERS.
For _y_.		
_e'_ - - -	- -	65 . 34 . 37
S t p - - -	- -	65 . 28 . 41
_y_ - - -		0 . 5 . 56
For correcting _m_ and _m'_.		
_m_ - - -	- -	20 . 4 . 43
_y_ + - - -	- -	0 . 5 . 56
_m_ correct. - -	- -	20 . 10 . 39
_m'_ - - -	- -	23 . 18 . 28
_y_ — - -	- -	0 . 5 . 56
_m'_ correct. -	- -	23 . 12 . 32
For t p.		
Sin. S t p - a. c.	0 . 0410635	
S p - -	9 . 9638681	0 . 92017
Sin. T' S P' - -	9 . 4784823	17 . 30 . 51
t p - - -	9 . 4834139	0 . 30438
For T P.		
t p - - -	9 . 4834139	
Sin. _m'_ - -	9 . 5955900	
L' - - -	0 . 4641410	
T P - - -	9 . 5431449	0 . 34926
For T″ P″.		
t p - - -	9 . 4834139	
Sin. _m_ - -	9 . 5377350	
L″ - - -	0 . 4641410	
T″ P″ - -	9 . 4852899	0 . 30570

G g

For P C.	LOGARITHMS.	NUMBERS.
T P - - -	9 . 5431449	$\overset{o}{17} . \overset{\prime}{51} . \overset{\prime\prime}{39}$
Tangent l - -	9 . 5081723	
P C - - -	9 . 0513172	0 . 11254

For P″ C″.		
T″ P″ - - -	9 . 4852899	$\overset{o}{23} . \overset{\prime}{43} . \overset{\prime\prime}{55}$
Tangent $l″$ - -	9 . 6430910	
	9 . 1283809	0 . 13439

For T S P.		
S T P = e - - -	- -	$\overset{o}{81} . \overset{\prime}{45} . \overset{\prime\prime}{54}$
Two remaining Angles -	- -	98 . 14 . 6
Half of them - -	- -	49 . 7 . 3
S T - - -	- -	1 . 00740
T P - - -	- -	0 . 34926
S T + T P - a. c.	9 . 8675290	1 . 35666
S T − T P - -	9 . 8183183	0 . 65814
Tangent - -	10 . 0626363	$\overset{o}{49} . \overset{\prime}{7} . \overset{\prime\prime}{3}$
Tangent - -	9 . 7484836	29 . 15 . 56
T S P - - -	- -	19 . 51 . 7

For T″ S P″.		
S T″ P″ = $e″$ - -	- -	$\overset{o}{46} . \overset{\prime}{9} . \overset{\prime\prime}{58}$
Two remaining Angles -	- -	133 . 50 . 2
Half of them - -	- -	66 . 55 . 1
S T″ - -	- -	1 . 00500
T″ P″ - -	- -	0 . 30570
S T″ + T″ P″ a. c.	9 . 8824967	1 . 31070
S T″ − T″ P″ - -	9 . 8446635	0 . 69930
Tangent - -	10 . 3704000	$\overset{o}{66} . \overset{\prime}{55} . \overset{\prime\prime}{1}$
Tangent - -	10 . 0975602	51 . 22 . 55
T″ S P″ - -	- -	15 . 32 . 6

For S P.			LOGARITHMS.	NUMBERS.
Sin. T S P	-	a. c.	0 . 4690460	19 . 51 . 7
T P	-	-	9 . 5431449	0 . 34926
Sin. *e*	-	-	9 . 9954987	81 . 45 . 54
S P	-	-	0 . 0076896	1 . 01784

For S P″.				
Sin. T″ S P″	-	a. c.	0 . 5721456	15 . 32 . 6
T″ P″	-	-	9 . 4852899	0 . 30570
Sin. *e″*	-	-	9 . 8581465	46 . 9 . 58
S P″	-	-	9 . 9155820	0 . 82355

For S C.				
S P	-	a. c.	9 . 9923104	
P C	-	-	9 . 0513172	
Tangent P S C	-	-	9 . 0436276	6 . 18 . 35
Sin. P S C	-	a. c.	0 . 9590091	
P C	-	-	9 . 0513172	
S C	-	-	0 . 0103263	1 . 02406

For S C″.				
S P″	-	a. c.	0 . 0844180	
P″ C″	-	-	9 . 1283809	
Tangent P″ S C″	-		9 . 2127989	9 . 16 . 18
Sin. P″ S C″	-	a. c.	0 . 7929141	
P″ C″	-	-	9 . 1283809	
S C″	-	-	9 . 9212950	0 . 83425

For P S P″.	LOGARITHMS.	NUMBERS.
T S T″ - -	- -	7 . 47 . 15
T″ S P″ - -	- -	15 . 32 . 6
		23 . 19 . 21
T S P - - -	- -	19 . 51 . 7
P S P″ - - -	- -	3 . 28 . 14
For P P″.		
P S P″ - -	- -	3 . 28 . 14
Two remaining Angles -	- -	176 . 31 . 46
Half of them, - -	- -	88 . 15 . 53
S P - - -	- -	1 . 01784
S P″ - - -	- -	0 . 82335
S P + S P″ - a. c.	9 . 7349014	1 . 84119
S P − S P″ - -	9 . 2888973	0 . 19449
Tangent - -	11 . 5186208	88 . 15 . 53
Tangent - -	10 . 5424195	73 . 59 . 50
S P P″ - -	- -	14 . 16 . 3
Sin. S P P″ - a. c.	0 . 6082724	
S P″ - -	9 . 9155820	
Sin. P S P″ -	8 . 7819944	
P P″ - -	9 . 3058488	0 . 20223
For C C″.		
P″ C″ - - -	- -	0 . 13439
P C - - -	- -	0 . 11254
P″ C″ − P C = I C″	- -	0 . 02185
P P″ - - a. c.	0 . 6941512	
I C″ - - -	8 . 3394514	
Tangent C″ C I - -	9 . 0336026	6 . 10 . 0
Sin. C″ C I - a. c.	0 . 9689170	
I C″ - - -	8 . 3394514	
C C″ - - -	9 . 3083684	0 . 203408

	LOGARITHMS.	NUMBERS.
Formation of the Equation $b c^2$ $-\frac{c^4}{12 b}$, and its comparison with a.		

For $b c^2$.	LOGARITHMS.	NUMBERS.
$c = $ C C″ - -	9 . 3083684	0 . 203408
S C - -	- -	1 . 02406
S C″ - -	- -	0 . 83425
$b = $ S C $+$ S C″ -	0 . 2691181	1 . 85831
c^2 - - -	8 . 6167368	
$b c^2$ - - -	8 . 8858549	0 . 0768874

For $12 b$.		
b - - -	0 . 2691181	
12. Constant - -	1 . 0791812	
$12 b$ - - -	1 . 3482993	

For $\frac{c^4}{12 b}$.		
$\frac{c^4}{12 b}$ - - -	7 . 2334736	
	1 . 3482993	
$\frac{c^4}{12 b}$ - - -	5 . 8851743	0 . 00007677

For $b c^2 - \frac{c^4}{12 b}$, and for comparing it with a.		
$b c^2$ - - -	- -	0 . 07688740
$\frac{c^4}{12 b}$ - - -	- -	0 . 00007677
$b c^2 - \frac{c^4}{12 b}$ - -	- -	0 . 07681063
a - - -	- -	0 . 07575370
$g +$ - - -	- -	0 . 0010569

H h

PART THE SECOND.

	NUMBERS.
In the Second Pofition	
T′ P′ was affumed -	0 . 303
Whence were deduced	
S C - - - -	1 . 023530
S C″ - - - -	0 . 835050
C C″ - - - -	0 . 202063

And the Equation $bc^2 - \dfrac{c^4}{12b}$ formed from thefe Quantities gives   -   -   0 . 0758086

$\dfrac{a}{g'} +$ -   -   -   - $\underline{0 . 0757537}$

0 . 0000549

PART THE THIRD.

From the Errors of the two former Pofitions,

	NUMBERS.
$g +$ - - -	0 . 0010569
and $g' +$ - - -	0 . 0000549

the Equation $\dfrac{g'b}{g-g'}$ was formed, and applied to S C, S C″, and C C″, as in the Example to the Graphical Procefs, Part 3d.

The Quantities thus corrected are

	NUMBERS.
S C - - - -	1 . 0235010
S C″ - - - -	0 . 8350938
C C″ - - - -	0 . 2019893

	NUMBERS.
With thefe the Equation $b c^2 - \frac{c^4}{12 b}$, was once more formed, and the refult was -	0 . 07575548
a - - - -	0 . 07575370
Error + - - -	0 . 00000178

This Error is fo fmall, that it is evident the Theory is extremely near the truth; and that the Quantities of the Second Pofition, cor-rected by the Formula $\frac{g' h}{g - g'}$, may be ufed for finding the Elements of the Orbit.

PART THE FOURTH.

Quantities to be ufed in the Determination of the Elements of the Comet's Orbit, corrected from thofe found in the Second Pofition, by the Equation $\frac{g' h}{g - g'}$.

				NUMBERS.
T P	-	-	-	0 . 3468340
T P''	-	-	-	0 . 3035590
P C	-	-	-	0 . 1117625
P'' C''	-	-	-	0 . 1334542
S P	-	-	-	1 . 0173760
S P''	-	-	-	0 . 8243630
P P''	-	-	-	0 . 2008220
S C	-	-	-	1 . 0235010
S C''	-	-	-	0 . 8350938
C C''	-	-	-	0 . 2019893
				° ' ''
S P P''	-	-	-	14 . 23 . 59
T'' S P''	-	-	-	15 . 24 . 15

Determination of the Elements of the Orbit of the Comet of 1769, by the Trigonometrical Method.

FOR THE LINE OF NODES.

For P″ R.				LOGARITHMS.	NUMBERS.
P″ C″	-	-	-	- -	0 . 1334542
P C	-	-	-	- -	0 . 1117625
P″ C″ — P C = C I		a. c.		1 . 6637064	0 . 0216917
P P″	-	-	-	9 . 3027896	0 . 2008220
P C	-	-	-	9 . 0482962	
P R	-	-	-	0 . 0147922	1 . 034650
P P″	-	-	-	- -	0 . 200822
P″ R	-	-	-	- -	1 . 235472

For P″ S R.					
P P″ S	-	-	-	- -	162 . 7 . 38
Two remaining Angles		-		- -	17 . 52 . 22
Half of them	-	-	-	- -	8 . 56 . 11
P″ R	-	-	-	- -	1 . 235472
S P″	-	-	-	- -	0 . 824363
P″ R + S P″	-	a. c.		9 . 6861676	2 . 059835
P″ R — S P″	-	-		9 . 6139569	0 . 411109
Tangent	-	-		9 . 1965811	8 . 56 . 11
Tangent	-	-		8 . 4967056	1 . 47 . 51
P″ S R		-		- -	10 . 44 . 2

For T″ S R.

				S.	°	′	″
T″ S P″	-	-	-		15 .	24 .	15
P″ S R	-	-	-		10 .	44 .	2
T″ S R	-	-	-		4 .	40 .	13

For the Longitude of the Node.

				S.	°	′	″
T″ S R	-	-	-		4 .	40 .	13
Earth's Longitude T″	-	-		11 .	20 .	29 .	20
Longitude of Defcending Node	-	-		11 .	25 .	9 .	33
Longitude of Afcending Node	-	-		5 .	25 .	9 .	33

For the INCLINATION of the ORBIT.

				LOGARITHMS.	NUMBERS.
S P″	-	-	-	9 . 9161185	0 . 824363
Sin. P″ S R	-	-		9 . 2700911	10 . 44 . 2
P″ C″	-	-	a. c.	0 . 8746678	0 . 133454
Cotangent	-	-		10 . 0608774	40 . 59 . 50
Inclination	-	-		- -	40 . 59 . 50

For the ANOMALY of S C, the GREATER RADIUS.

				LOGARITHMS.	NUMBERS.
S C	-	-	-	- -	1 . 023501
S C″	-	-	÷	- -	0 . 835093
C C″	-	-	-	- -	0 . 201988
Sum	-	-	-	- -	2 . 060582
½ Sum	-	-	-	- -	1 . 030291
½ Sum — S C	-	-		7 . 8318698	0 . 006790
½ Sum — C C″	-	-		9 . 9181893	0 . 828303
S C	-	-	a. c.	9 . 9899118	
C C″	-	-	a. c.	0 . 6946744	
Sum	-	-	-	8 . 4346453	
½ Sum. Sin.	-	-		9 . 2178226	9 . 29 . 37
C″ C S	-	-		- -	18 . 59 . 14

S C — S C″		-		9 . 2750995	0 . 188408
C C″		-	a. c.	0 . 6946744	
Cof. B C C″	-	-		9 . 9697739	21 . 7 . 46
C″ C S	-	-		- -	18 . 59 . 14
B C S	-	-		- -	40 . 7 . 0
Anomaly of S C	-	-		- -	139 . 53 ⸱ 0

For the PERIHELION DISTANCE.

½ Anomaly Cof.	-	-	9 . 5352646	69 . 56 . 30
			2	
Cof.² of ½ Anomaly		-	9 . 0705292	
S C		-	0 . 0100882	
Perihelion Diftance		-	9 . 0806174	0 . 1203975

For the TIME of ARRIVAL at the PERIHELION.

		LOGARITHMS.	NUMBERS.
Perihelion Diſtance	-	9 . 0806174	
Its Half	- -	9 . 5403087	
Perihelion Diſtance $^{\frac{3}{2}}$	-	8 . 6209261	
Days of Anomaly of S C	-	2 . 8966478	788,2206
			D.
Days from Perihelion	-	1 . 5175739	32 . 92865
Epoch. September	-	- -	4 . 5833
Arrival at Perihelion	October	- -	7 . 51195
			D. H. M. S.
or October	-	- -	7 . 12 . 17 . 13

For the PLACE of the PERIHELION on the ORBIT.

		LOGARITHMS.	NUMBERS.
			° ′ ″
Tangent of P″ S R	-	9 . 2777573	10 . 44 . 2
Coſine of Inclination	a. c.	0 . 1222018	40 . 59 . 50
Tangent of C″ S R	-	9 . 3999591	14 . 5 . 57
For the Anomaly of S C″	-		D.
Arrival at Perihelion	October	- -	7 . 51195
Time of S C″ -	September	- -	12 . 58330
Days -	-	1 . 3966814	24 . 92765
Perihelion Diſtance $^{\frac{3}{2}}$	-	8 . 6209261	
Days	- -	2 . 7757553	596 . 69
			° ′ ″
Anomaly of 596 . 69 Days = C″ S V	- -	135 . 23 . 27
C″ S R - -	- -	14 . 5 . 57
R S V - -	- -	149 . 29 . 24
Deſcending Node -	- -	355 . 9 . 33
Place of Perihelion -	- -	144 . 38 . 57
			S. ° ′ ″
or - -	- -	4 . 24 . 38 . 57
Motion -	- -	Direct.

Elements of the Orbit of the Comet of 1769, *as determined by the Trigonometrical Method.*

		S.	°	′	″
Longitude of the Afcending Node	- -	5	25	9	33
Inclination of the Orbit	- - -		40	59	50
Perihelion Diftance	- - -	0	. 1203975		
Its Logarithm	- - -	9	. 0806174		
		D.	H.	M.	S.
Arrival at the Perihelion	- October	7	12	17	13
		S.	°	′	″
Longitude of the Perihelion on the Orbit	-	4	24	38	57
Motion	- - -		Direct.		

Elements determined by PROSPERIN, *fee* Comêtographie, *Vol. 2, Page* 247.

		S.	°	′	″
Longitude of Afcending Node	- -	5	25	6	33
Inclination of the Orbit	- -		40	48	49
Perihelion Diftance	- - -	0	. 12272		
Its Logarithm	- - -	9	. 0889150		
		D.	H.	M.	S.
Arrival at the Perihelion	- October	7	13	58	40
		S.	°	′	″
Longitude of the Perihelion	- -	4	24	11	7
Motion	- -		Direct.		

CONCLUSION.

§ 1. **I**T has been already obferved, that the Elements obtained by the method defcribed in the preceding pages, can only be confidered as approximate, and muft be further corrected by diftant obfervations. This is equally true of every other method founded on obfervations made at fmall intervals from each other; and it is evident that it cannot be otherwife; as the effect of every little error in obfervation, or want of abfolute precifion in calculation, though infenfible on a fhort interval, muft go on increafing, when obfervations made at diftant times are compared with the refults of the former computation. The exact Elements of the Planets themfelves are only deducible from diftant obfervations. Much more muft it be neceffary to have recourfe to the fame method for Comets, whofe Orbits are of a figure difficult to determine with accuracy; and whofe Difk or Nucleus is always fo ill defined as to render great accuracy of obfervation impoffible; and very often is totally undiftinguifhable, from the Haze or Coma which furrounds it.

§ 2. Having therefore obtained approximate Elements of the Comet's Orbit, fufficient to give its apparent Path in the Heavens, during the time of its appearance, without any enormous error; which by the method above defcribed will almoft always be the cafe; and by Mr. de la Place's method given hereafter, will be certainly attained; it will fcarce ever be worth while to

K k

go any further, till from a careful inveſtigation of the whole body of obſervations made on the Comet, three may be ſelected, from which the correct Elements of the Comet's Orbit may be deduced.

§ 3. Various methods have been propoſed for this operation; all founded on the principle of Falſe Poſition; in which, by finding the changes of errors induced by ſmall changes in two of the Elemènts, data are obtained for ſuch changes as ſhall deſtroy the errors. The difference of the methods does therefore chiefly conſiſt in the different Elements which are varied in the Poſitions. Boſcovich changes the Longitude of the Aſcending Node, and Inclination of the Orbit. Mr. de la Place, whoſe method is now moſt generally followed, makes the variation fall on the Perihelion Diſtance, and Time of Arrival at the Perihelion. As the latter method is given in the utmoſt detail in the Second part of this work, it was judged ſuperfluous to enter into any detail of Boſcovich's method. As however in a ſhort Treatiſe, in the 5th Vol. of the Collection of Tracts, printed at Baſſano, whence the method at large was taken, Page 383; he gives a very elegant and ſimple method for deducing the Heliocentric Longitude and Latitude, and Radius Vector of a Comet, from its obſerved Geocentric Place, the Poſition of the Line of Nodes, and the Inclination of the Orbit being given; this method, which would not have entered properly in any other place, will be given here as a concluſion to this part of the work.

§ 4. Previous however to the entering on any method for the final determination of the Elements of the Orbit of a Comet; it may not be amiſs to obſerve, that although, as has been before ſtated, it is not often worth while to enter into the laborious computation of accurate Elements, till the final rectification is undertaken, yet infpection of the Figure prepared for the approximate Elements of the Comet, will often guide us to a guefs at fuch corrections as will much diminifh the errors of the Elements. In the cafe of the Comet of 1769, which ferved as an Example to the method of Bofcovich, it is obvious that the Comet on the 3d of November, was by the Elements of Figure 16, too far advanced in its Orbit; and as the perpendicular diftances of the Comet from the Plane of the Ecliptic were pretty rapidly increafing, it is evident, that fuch a change in the Elements as carried it nearer the Sun, would alfo diminifh its diftance from the Ecliptic, and of courfe its Geocentric Latitude. In fact, the principal error in thefe Elements, lies in the Perihelion diftance being too fmall; (an error which they have in common with all others determined from obfervations made in the firft branch of the Orbit); if therefore by guefs, the Perihelion diftance fhould be a little increafed, leaving the other Elements untouched; the errors in Longitude and Latitude would be very much diminifhed. Habit, and an attentive confideration of the Pofitions of the Earth and Comet, with refpect to each other, and to the Sun; will in moft cafes lead to fuch corrections of the Elements as will much fhorten the future work when the precife Elements are fought.

§ 5. It now remains to give the method propofed by Bofcovich, for obtaining the Heliocentric Place of a Comet. It is as follows.

PROBLEM.

THE Longitude of the Node, and Inclination of the Orbit of a Comet being given; to find the Heliocentric Longitude and Latitude, and Radius Vector, from the Geocentric Longitude and Latitude obſerved.

§ 6. Let S, (Figure 17), be the Sun ; T, the Place of the Earth ; C, the Place of the Comet in its Orbit; P, the Place of the Comet reduced to the Ecliptic ; and S N, the Line of Nodes. From P, draw P D, perpendicular to the Line of Nodes; and ſuppoſe the Triangle P D C, to be perpendicular to the Plane of the Ecliptic.

§ 7. The Angle P D C, will be the Inclination of the Orbit ; P T C, the Geocentric Latitude ; whoſe Tangents are, $\frac{PC}{PD}$, and $\frac{PC}{PT}$: we therefore have, $\frac{PT}{PD}, = \frac{\text{Tang. Inclin.}}{\text{Tang. Lat.}}$. T P, is the direction of the given Longitude ; and D P, that of a Longitude, 90° diſtant from that of the Node ; the difference of theſe Longitudes gives the Angle T P D. And the Ratio of the two Sides containing it, P T, and P D, being known, they being ſhewn above to be the Tangents of the Geocentric Latitude, and Inclination of the Orbit; the Angles T D P, P T D, may be thence found.

§ 8. The Angle T D S, is equal to the difference of the Angles T D P, and S D P, = 90°; or their ſum, according to the different cafes; the Angle T S D, is equal to the difference of the Heliocentric Longitude of the Earth, and that of the Node. Having therefore in the Triangle S T D, two of the Angles; the third, S T D, is found.

§ 9. In the fame Triangle; the Side S T, is given; whence we obtain, $SD = \dfrac{ST \times Sin.\,STD}{Sin.\,TDS}$; $TD = \dfrac{ST \times Sin.\,TSD}{Sin.\,TDS}$;

$DP = \dfrac{TD \times Sin.\,P\dot{T}D}{Sin.\,TPD}$; $= \dfrac{ST \times Sin.\,TSD \times Sin.\,PTD}{Sin.\,TDS \times Sin.\,TPD}$. We

therefore have, Tang. $DSP = \dfrac{DP}{SD} = \dfrac{Sin.\,TSD \times Sin.\,PTD}{Sin.\,TPD \times Sin.\,STD}$.

Thus the Angle D S P, is found without the neceffity of finding any Side; and from D S P, D S C, may be found thus:

Tang. $DSC = \dfrac{Tang.\,DSP \times DC}{DP} = \dfrac{Tang.\,DSP}{Cof.\,Inclin.}$.

§ 10. Having the Longitude of the Node, S N; by adding to, or fubtracting from it, the Angle D S P, we find the Heliocentric Longitude of the Comet in the Ecliptic: and in the fame manner, with the Angle D S C, will its Longitude in its Orbit be found.

§ 11. The Curtate Diftance S P, will be $= \dfrac{SD}{Cof.\,DSP} =$

$\dfrac{ST \times Sin.\,STD}{Sin.\,TDS \times Cof.\,DSP}$; and the Radius Vector $SC = \dfrac{SD}{Cof.\,DSC}$

$= \dfrac{ST \times Sin.\,STD}{Sin.\,TDS \times Cof.\,DSC}$. Thus by a fimple and eafy calculation we have obtained the Comet's Heliocentric Longitude in the Ecliptic, and on the Orbit, the Curtate Diftance from the Sun, and the Radius Vector.

L l

§ 12. From the fame data we have,

$$TP = \frac{TD \times Sin. \, TDP}{Sin. \, TPD} = \frac{ST \times Sin. \, TSD \times Sin. \, TDP}{Sin. \, TDS \times Sin \, TPD};$$

$$TC = \frac{TP}{Cof. \, PTC} = \frac{TP}{Cof. \, Lat. \, Geoc.}; \text{ and,}$$

$$Sin. \, PSC = \frac{PC}{SC} = \frac{TP \times Tang. \, Lat. \, Geoc.}{SC}. \qquad \text{Thefe three}$$

Equations give the Comet's Curtate Diftance from the Earth,
its entire Diftance, and its Heliocentric Latitude; and if
the latter alone is wanted, without finding the Lines T P,
and S C; their values found as above, may be fubftituted in
the Formula for P S C, which will then become, Sin. P S C =

$$\frac{Sin. \, TSD \times Sin. \, TDP \times Cof. \, DSC \times Tang. \, Lat. \, Geoc.}{Sin. \, TPD \times Sin. \, STD.}$$

The operation therefore for the Heliocentric Longitude and
Latitude, after finding the three Angles of the Triangle P D T,
is comprized in three Equations; and by a fourth the Radius
Vector is found. And this is performed with great eafe and
fimplicity by the following Graphical method.

§ 13. Let S, (Figure 18), be the Sun; T, the Earth; S T,
the Earth's Diftance from the Sun; S N, the Line of Nodes;
S T E, the Geocentric Longitude of the Comet, or its Elongation from the Sun, the Line T E, being prolonged indefinitely. On T E, fet off T A, = to the Earth's Mean Diftance from the Sun, taken as 1,0000. Find the value in num-
bers of the Fraction $\frac{Tang. \, Lat.}{Tang. \, Inclin.}$, and having from A, drawn
an indefinite Line A B, perpendicular to the Line of Nodes;
fet off this value, on the Line A B, from A, to B. Through
T, and B, draw a Line cutting the Line of Nodes in D, and

from D, draw D P, parallel to A B, and of courfe perpendicular to the Line of Nodes, meeting T E, in P. With D P, as
Radius on the Line of Secants on the Sector, take the Secant
of the Inclination of the Comet's Orbit, and fet it off from D,
on the Line D P, prolonged to C. Then P, will be the projected place of the Comet in its Orbit; and C, its place in the
Orbit, applied to the Plane of the Ecliptic:

For $\dfrac{P D}{P T} = \dfrac{A B}{A T} = \dfrac{\text{Tang. Lat.}}{\text{Tang. Inclin.}}$; and

$D C = D P \times \text{Sec. Inclin.}$ We have alfo $D C = \dfrac{D P}{\text{Cof. Inclin.}}$

§ 14. A Line drawn from S, through P, gives the Heliocentric Longitude of the Comet; S C, is the Radius Vector, and if
with D C, for Radius on the Line of Sines on the Sector, the
Sine of the Inclination be found, then will that Sine be the
Tangent of the Heliocentric Latitude of the Comet, to the Radius S P; or its Sine to the Radius S C.

§ 15. This Graphical method, befides other ufes, may be extremely convenient for finding the Heliocentric Longitude and
Latitude, of the Extremity of a Comet's Tail ; and from thence
its deviation from oppofition to the Sun; that deviation being
fuppofed to be ever in the Plane of the Comet's Orbit. For if
there be no deviation, the Longitude and Latitude thus found
will be the fame as thofe of the Comet itfelf; and if there be,
the deviation will be the Hypothenufe of the Right-angled
Spherical Triangle, whofe Sides are the differences in Longitude and Latitude of the Tail, from the Head of the Comet.
But of this probably more hereafter.

GENERAL METHOD for DETERMINING the ORBITS of COMETS.

BY MR. DE LA PLACE.

§ 1. THIS method will be divided into Two Parts. The Firſt, will give a method for obtaining an approximation to the Perihelion Diſtance, and Time of the Comet's Arrival at the Perihelion; the Second Part will contain the correction of the approximate Perihelion Diſtance, and Time of Perihelion, from more diſtant obſervations; and a determination of the remaining Elements of the Comet's Orbit.

PART I

Determination of the Approximate Perihelion Diſtance, and Time of the Comet's Arrival at the Perihelion.

§ 2. Chuſe Three, Four, or Five Obſervations of the Comet, as nearly equidiſtant from each other as poſſible; and which, for the convenience of calculation, had better be reduced to the ſame hour; if four Obſervations are uſed, the Arch of the Comet's Geocentric Motion may be about 30°; if five Obſervations be taken, the Arch may be of 36° or 40°. In general, the more Obſervations are uſed, the greater may be the Arch of the Comet's Motion; and by that means the Influence of the Errors of Obſervation on the Operation, will be diminiſhed.

M m

§ 3. Now let ς, ς', ς'', ς''', &c. be the Longitudes of the Comet; γ, γ', γ'', γ''', &c. be its Latitudes; the Northern Latitudes being suppofed $+$, the Southern, $-$; divide the difference of Longitude betw.een the firft and fecond Obfervation, $= \varsigma' - \varsigma$; by the time in days, elapfed between the firft and fecond Obfervation, $= \tau$; and call $\frac{\varsigma' - \varsigma}{\tau}$, $\delta \varsigma$. In the fame manner divide $\varsigma'' - \varsigma'$, by the time in days, elapfed between the fecond and third Obfervation, $= \tau'$; and call $\frac{\varsigma'' - \varsigma'}{\tau'}$, $\delta \varsigma'$, and fo on, calling this feries of Quotients $\delta \varsigma$, $\delta \varsigma'$, $\delta \varsigma''$, $\delta \varsigma'''$, &c.

§ 4. Divide the difference of $\delta \varsigma$, and $\delta \varsigma'$, by the time elapfed between the firft and third Obfervation, $= \tau + \tau'$, and call the Quotient $\frac{\delta \varsigma' - \delta \varsigma}{\tau + \tau'}$, $\delta^2 \varsigma$. In the fame manner divide $\delta \varsigma'' - \delta \varsigma'$, by the number of days that feparate the fecond and fourth Obfervation, $= \tau' + \tau''$, and call the Quotient $\frac{\delta \varsigma'' - \delta \varsigma'}{\tau' + \tau''}$, $\delta^2 \varsigma'$, and fo on; calling this feries of Quotients $\delta^2 \varsigma$, $\delta^2 \varsigma'$, $\delta^2 \varsigma''$, $\delta \varsigma'''$, &c.

§ 5. Divide the difference, $\delta^2 \varsigma' - \delta^2 \varsigma$, by the number of days elapfed between the firft and fourth Obfervation, $= \tau + \tau' + \tau''$. And call the Quotient $\frac{\delta^2 \varsigma' - \delta^2 \varsigma}{\tau + \tau' + \tau''}$, $\delta^3 \varsigma$, and fo on; calling this feries of Quotients $\delta^3 \varsigma$, $\delta^3 \varsigma'$, &c.

§ 6. Proceed in the fame manner with thefe laft Quotients, and fo on, till a Quotient is obtained $\delta^{n-1} \varsigma$; n being the number of Obfervations employed.

§ 7. Chufe for Epoch, a Time equidiftant, or nearly fo, from the two extreme Obfervations, and call the number of days which the Epoch is diftant from each Obfervation; i, i', i'', &c. i, i', &c. being fuppofed Negative, for thofe Obfervations which precede the Epoch. The Longitude of the Comet for a moment diftant from the Epoch, a number of days, which call z, will be given by this Formula.

$$\mathfrak{C} - i\,\delta\mathfrak{C}. + i\,i'\,\delta^2\mathfrak{C} - i\,i'\,i''\;\delta^3\mathfrak{C} + i\,i'\,i''\,i'''\;\delta^4\,\mathfrak{C}, - \&c.$$

$$+ z \left\{ \begin{array}{l} \delta\mathfrak{C} - (i + i')\,\delta^2\,\mathfrak{C}. + (i\,i' + i\,i'' + i'\,i'')\,\delta^3\,\mathfrak{C}. - (i\,i'\,i'' \\ + i\,i'\,i''' + i\,i''\,i''' + i'\,i''\,i''')\,\delta^4\,\mathfrak{C} + \&c. \end{array} \right\}$$

$$+ z^2 \left\{ \begin{array}{l} \delta^2\mathfrak{C} - (i + i' + i'')\,\delta^3\,\mathfrak{C} + (i\,i' + i\,i'' + i\,i''' + i'\,i'' \\ + i'\,i''' + i''\,i''')\,\delta^4\,\mathfrak{C} - \&c. \end{array} \right\}$$

§ 8. In the Firft Member of this Formula, which is independent of z, the Coefficients of $- \delta\mathfrak{C} + \delta^2\mathfrak{C} - \delta^3\mathfrak{C}$, &c. are firft, the time between the Epoch and the firft Obfervation, $= i$: Secondly, the Product of the time between the Epoch and firft Obfervation, and the time between the Epoch and fecond Obfervation, $= i\,i'$: Thirdly, the Product of the three times, i, i', i'', and fo on.

§ 9. In the Second Member, which is multiplied by z; the Coefficients of $- \delta^2\mathfrak{C}, + \delta^3\mathfrak{C} - \delta^4\mathfrak{C}$, &c. are Firft, the Sum of the Time between the Epoch and the firft Obfervation, and the Time between the Epoch, and Second Obfervation: Secondly, the Sum of the Products of i, i', and i'', multiplied two by two: Thirdly, the Sum of the Products of i, i', i'', and i''', multiplied three by three; and fo on.

§ 10. In the Third Member, which is multiplied by z^2, the Coefficients of $- \delta^3\mathfrak{C}, + \delta^4\mathfrak{C}, - \delta^5\mathfrak{C}$, &c. are, Firft, the Sum of the Time, between the Epoch, and firft Obfervation; the Time between the Epoch, and fecond Obfervation; and the Time between the Epoch, and third Obſervation : Secondly, the Sum of the Products of the four times i, i', i'', and i''', multiplied two by two : Thirdly, the Sum of the Products of the five times, i, i', i'', i''', and i'''', multiplied three by three, &c.

§ 11. If five Obſervations are uſed, and the Epoch is fixed at the third Obſervation, the Formula is much ſimplified ; for i'' being in that caſe = 0 ; all the terms of the foregoing Formula in which i'' is found, and all the Multiples in which i'' is a factor, are reduced to Zero ; and the Member independent of z, viz, $c — i\,\delta c + i\,i'\,\delta^2 c — i\,i'\,i''\,\delta^3 c + i\,i'\,i''\,i'''\,\delta^4 c$, be‑ comes $c — i\,\delta c + i\,i'\,\delta^2 c$; (the other Terms vaniſhing, as con‑ taining i'') ; which is equal to the Longitude of the third Ob‑ ſervation choſen for the Epoch. It therefore becomes ſuper‑ fluous to calculate it ; and the Formula is reduced to

$$c'' + z \left\{ \delta c — (i + i')\,\delta^2 c + (i\,i')\,\delta^3 c — (i\,i'\,i''')\,\delta^4 c. \right\}$$
$$+ z^2 \left\{ \delta^2 c — (i + i')\,\delta^3 c + (i\,i' + i\,i''' + i'\,i''')\,\delta^4 c. \right\}$$

§ 12. If the intervals of time between the five Obſervations are equal, and the Epoch is fixed at the third Obſervation ; the Formula is ſtill farther ſimplified ; for it becomes

$$c'' + z \left(\frac{c — c''''}{12\,i} — \frac{2\,c''' — 2\,c'}{3\,i} \right) + z^2 \left(\frac{2\,c' + 2\,c'''}{3\,i^2} — \frac{5\,c''}{4\,i^2} — \frac{c''' + c}{24\,i^2} \right). \,*$$

and all the preceding calculation of the δc, &c. becomes un‑ neceſſary ; this Formula not comprehending any of them.

* It is to be obſerved that in this Formula, i, means ſimply the Interval of Time between the Obſervations ; and not, as in the former Formula, the Interval between the firſt Obſervation and the Epoch. In this For‑ mula i, is always Poſitive.

§ 13. The whole Operation for the Latitudes, is precisely the same as that for the Longitudes; and the Formula the same; substituting in them the Letter γ, for ε. The Northern Latitudes, as before observed, being supposed Positive, and the Southern, Negative.

§ 14. The foregoing Formulæ, for finding a Longitude and Latitude of the Comet, at a time a little preceding or following the Time of the Epoch, are only useful for that purpose when the Operation is performed on the Right Ascensions and Declinations of the Comet; as Mr. De La Place at first proposed; in which case ε, was the Right Ascension of the Comet, and γ, its Declination: and it was then necessary to compute by the Formulæ two places of the Comet, distant from the Epoch by the number z, of days. But if the Longitudes and Latitudes of the Comet are made the basis of the Operation, as is directed above, which is by much the best way; these Formulæ are of use only for the formation of the four quantities, a, b, h, and l, from the Coefficients of z, and z^2; in the following manner.

§ 15. * 1. In the Formula for the Longitudes, reduce into Seconds the Coefficient of z; and from the Logarithm of that Number, subtract the constant Logarithm 3,5500081. The remainder is the Logarithm of a Number, which call, a.

* It is necessary here to take notice, that those who consult Mr. Pingrè, will find a fault of the greatest consequence in the Rule he gives (Page 328), for the formation of the Quantities a, b, h, and l; or as he calls them, β, γ, μ, and ν. The Quantities themselves in his book are rightly formed. It is the Rule only that is erroneously given.

N n

2. In the fame Formula, reduce into Seconds the Coefficient of z^2; and from the Logarithm of twice that Number, fubtract the conftant Logarithm 1,7855911. The remainder is the Logarithm of a Number, which call, b.

3. In the Formula for the Latitudes, reduce into Seconds the Coefficient of z; and from the Logarithm of that Number, fubtract the conftant Logarithm 3.5500081. The remainder is the Logarithm of a Number, which call, h.

4. In the fame Formula, reduce into Seconds the Coefficient of z^2; and from the Logarithm of twice that Number, fubtract the conftant Logarithm 1.7855911. The remainder is the Logarithm of a Number, which call, l.

In the formation of thefe Numbers, the utmoft precifion muft be ufed; as on their accuracy the fuccefs of the whole Operation, does in a great meafure, depend.

§ 16. Let us now call the Longitude of the Comet, at the moment chofen for Epoch, u; its Latitude for that moment, θ; (they are the fame, if the Epoch is taken, as is moft advifable when it can be done, at the moment of the middle Obfervation of five; as ε'', and γ''); find from the Solar Tables, or Ephemeris, the Longitude of the Earth for the moment of the Epoch, which call, A; the Radius Vector of the Earth for the fame moment, which call R, and the Radius Vector of the Earth for a Longitude 90° forwarder in the Ecliptic than its place at the Epoch $= A + 90°$, which call R'.

§ 17. With thefe Quantities the refearches of M:. De La Place, (for the detail of which fee the *Memoires de L'Academie des Sciences*, for the Year 1780), give him the four following Equations.

EQUATION 1.

$$r^2 = \frac{x^2}{Cof.\ \theta^2} + 2\,R\,x\,Cof.\ (A - \alpha) + R^2.\ *$$

EQUATION 2.

$$y = \frac{Sin.\ (A - \alpha)}{2\,a\,R^2} - \frac{R\,Sin.\ (A - \alpha)}{2\,a\,r^3} - \frac{b\,x}{2\,a}.$$

EQUATION 3.

$$o = y^2 + a^2\,x^2 + \left(y\,Tang.\ \theta + \frac{b\,x}{Cof.\ \theta^2} \right)^2$$

$$+ 2\,y\,\left\{ (R' - 1)\ Cof.\ (A - \alpha) - \frac{Sin.\ (A - \alpha)}{R} \right\}$$

$$+ 2\,a\,x\,\left\{ (R' - 1)\ Sin.\ (A - \alpha) + \frac{Cof.\ (A - \alpha)}{R} \right\}$$

$$+ \frac{1}{R^2} - \frac{2}{r}.$$

EQUATION 4.

$$y = -\,x\,\left(b\,Tang.\ \theta + \frac{1}{2\,b} + \frac{a^2\,Sin.\ \theta\,Cof.\ \theta}{2\,b} \right).$$

$$+ \frac{R\,Sin.\ \theta\,Cof.\ \theta.}{2\,b}\ Cof.\ (A - \alpha)\ \left(\frac{1}{r^3} - \frac{1}{R^3} \right).$$

In thefe Equations, x, is the Curtate Diftance of the Comet from the Earth; which is to be taken as near the truth as may be, either by guefs from the fize and motion of the Comet, or

* In Mr. Pingrè, (Page 330), there is a falfe Print of great importance in the firft Member of the firft Equation, which he has printed r, inftead of r^2.

by the methods defcribed before in this work ; *r*, is the Radius
Vector of the Comet ; and *y*, is a Ratio of the Elements of Diftance, to the Elements of Time.

§ 18. In the folution of thefe Equations, it will be moft com-
modious to employ the Logarithms of the Coefficients. Having
made a firft fuppofition for *x*, the values of *r*, and *y*, muft be
deduced from it, by the Equations 1, and 2. Thefe values
muft then be ufed in the Solution of Equation 3, and if the
Equation comes out $= 0$; it is a proof that the value of *x*, was
rightly chofen ; if the remainder is Pofitive, the value of *x*,
muft be diminifhed. If the remainder is Negative, the value of
x, muft be increafed, and the Equations 1, and 2, again folved ;
which operation will not be long, as all the Coefficients of *x*, *r*,
and *y*, remain unchanged. A few trials will give the true value
of *x*.

§ 19. But as thefe Equations which arife from the decom-
pofition of Equations of an higher order, may, ftrictly fpeak-
ing, be fufceptible of more than one real value ; in order to af-
certain whether the value of *x*, thus obtained, is the true one,
the Equation 4, muft be ufed ; and if the value of *y*, given by
this Equation, is the fame, or nearly the fame as the value of *y*,
found by the Equation 2 ; (and this Mr. Pingrè thinks will
probably be always the cafe ;) the value of *x*, as before found,
is certainly the true one.

§ 20. If *l*, is greater than *b*, the Equation 4, fhould be employed to obtain the value of *y*, inftead of the Equation 2 ; and
then having brought the refult of Equation 3, to $= 0$, as before ; the Equation 2, will ferve as a verification, inftead of
Equation 4.

§ 21. Having obtained the values of x, y, and r; the Perihelion Diftance will be given by the following Equations.

EQUATION 1.

$$P = \frac{x}{Cof.\ \theta^2}\ (y + b\ x\ \mathit{Tang}.\ \theta) + R\ y\ \mathit{Cof}.\ (A - \alpha)$$

$$+\ x\ \left\{ (R' - 1)\ \mathit{Cof}.\ (A - \alpha) - \frac{\mathit{Sin}.\ (A - \alpha)}{R} \right\}$$

$$+\ R\ a\ x\ \mathit{Sin}.\ (A - \alpha) + R\ (R' - 1).$$

EQUATION 2.

$$D = r - \tfrac{1}{2} P^2$$

D being the Perihelion Diftance of the Comet.

§ 22. The Cofine of the Anomaly of the Comet, which call v, is given by the following Equation.

$$Cof.\ v = \frac{2\ D}{r} - 1.$$

Whence by the Table of the Parabolic Motion of Comets, the time employed by the Comet, to move the Angle v, will be found; and to obtain the time of the Comet's paffing the Perihelion, this time muft be added to the time of the Epoch, if P was Negative, and fubtracted from the Epoch, if P was Affirmative; as in the firft cafe, the Comet was approaching to the Perihelion, and in the latter cafe, receding from it.

O o

§ 23. If it is wished to obtain the rest of the Elements of the Comet's Orbit, from the approximate Perihelion Distance, and time of the Arrival at the Perihelion, here found; the method given in the Second Part, for finding them from the Corrected Perihelion Distance, and Arrival at the Perihelion, is of course equally applicable to these; and the Elements so found will be fully sufficient to predict the Comet's Motion, so as to find it after a few days interruption of the Observations.

§ 24. In the course of this whole Operation, a constant attention must be had to the Algebraic Signs, which obviously are of absolute necessity to give a true result. It may not therefore be useless here to observe, that if the Geocentric Motion of the Comet is Retrograde, the $\delta \epsilon$ must have the Sign —, being the first differences of Quantities which are diminishing; if these Negative $\delta \epsilon$ also diminish; the $\delta^2 \epsilon$ become Positive, because if — 5 be subtracted from — 3, the result is + 2. If the $\delta \epsilon$ being Negative, go on increasing, the $\delta^2 \epsilon$ will also be Negative; as if — 5 be subtracted from — 7, the remainder is — 2.

§ 25. With respect to the Latitudes, they are ever supposed in Algebraical Formulæ, to be Positive when they are Northern, because their Sines, Cosines, and Tangents, are then Positive; and South Latitudes are supposed Negative, their Sines, &c. being then Negative. If therefore the Latitudes are Northern, and increasing, the $\delta \gamma$ are Positive; and will have the Sign +. If the Latitudes are Northern, decreasing, the $\delta \gamma$ are Negative; if the Latitudes are Southern, increasing, the $\delta \gamma$ are Negative; if they are South decreasing, the $\delta \gamma$ are Positive; in consequence of the Rule of Algebraic Signs. The $\delta^2 \gamma$ will be affected according to the progreffion of the $\delta \gamma$, and the fame for the $\delta^3 \gamma$.

§ 26. Equal attention muft be given throughout the evaluation of the Formulæ, to the Values of the Sines, Cofines, and Tangents; which it may not be ufelefs to give here.

Arc.	Sine.	Cofine.	Tangent.
At 0° or 360°	— 0	+ 1	— 0
From 0° to 90	+	+	+
At - 90	+ 1	+ 0	+ Infinite.
From 90 to 180	+	—	—
At - 180	+ 0	— 1	— 0
From 180 to 270	—	—	+
At - 270	— 1	— 0	+ Infinite.
From 270 to 360	—	+	—

Thofe Values which have 1 affixed to the Sign, are equal to Radius. Thofe with 0, are nothing. When the Sines, &c. are raifed to the Second Power, or Squared, they are evidently always Pofitive, as it is well known that $- \times - = +$. And fo of the higher even Powers; the Cubes and other odd Powers are Negative, if the Root is Negative.

PART II.

Determination of the true Elements of the Orbit, from the Approximate Perihelion Diſtance, and Time of Arrival at the Perihelion, before found.

§ 27. HAVING by the foregoing methods obtained an Approximation to the Perihelion Diſtance of the Comet, and the time of its paſſage through that point; in order to obtain the exact Elements of the Orbit, three diſtant Obſervations muſt be choſen. By the Perihelion Diſtance and Time of Arrival at the Perihelion, found above; the three true Anomalies of the Comet, and the three Radii Vectores correſponding to the times of the three Obſervations choſen, muſt be computed. This is done by the method given in the uſe of the General Table of the Parabola. Theſe Anomalies we call v, v', and v''; and if they are on different Sides of the Perihelion, they muſt have different Signs; thoſe preceding the Perihelion, being —, and thoſe following it +. Let r, r', and r'', be the three Radii Vectores of the Comet. The Angle between r, and r', will be obtained, by ſubtracting the firſt Anomaly from the ſecond; and the Angle between r and r'', by ſubtracting the firſt Anomaly from the third.* Let U be the firſt of theſe Angles; and U' the ſecond.

* If the Anomalies are on oppoſite Sides of the Perihelion, they muſt be added.

§ 28. Let u, u', and u'', be the three obferved Geocentric Longitudes of the Comet; θ, θ', and θ'', its three Geocentric Latitudes; ε, ε', and ε'', the three Elongations of the Comet; T, T', and T'', the three correfponding Longitudes of the Earth; R, R', and R'', the three Radii Vectores of the Earth, or its Diftances from the Sun; $\mathfrak{E}$, $\mathfrak{E}'$, and $\mathfrak{E}''$, the three Helio-centric Longitudes of the Comet; and ϖ, ϖ', and ϖ'', its He-liocentric Latitudes.

§ 29. Now in Figure 19, let S, be the Sun; T, be the Place of the Earth; C, the true Place of the Comet; P, its Place projected on the Plane of the Ecliptic. Then in the four Tri-angles, S T P, S T C, S P C, and T P C; we have given, the Side S T $= R$; the Angle S T P, $=$ the Elongation of the Comet; the Angle P T C, $=$ the Geocentric Latitude of the Comet; the Angles C P S, and C P T, which are Right Angles; and we take the Side S C $= r$, from the Approximate Elements above. Then the Heliocentric Longitude, and La-titnde of the Comet, are found as follows.

1. $Cof.\ \varepsilon.\ Cof.\ \theta.$ - $= Cof.\ S T C.$

2. $\dfrac{Sin.\ S T C.\ R}{r}$ - $= Sin.\ T C S.$

3. $180° - T C S - S T C = T S C.$

4. $\dfrac{Sin.\ \theta.\ Sin.\ T S C}{Sin.\ S T C}$ - $= Sin.\ \varpi.$

5. $\dfrac{Cof.\ T S C}{Cof.\ \varpi}$ - $= Cof.\ T S P.$*

* The Triangle, No. 1, is conftant; its given Sides being thofe found by obfervation. The Numerator of Triangle, No. 2, Sin. S T C $\times$ R,

P p

The Angle T S P, is the Heliocentric Angle between the Earth and Comet; the Heliocentric Longitude of the Comet, is therefore found by adding or subtracting that Angle from the Earth's Place, = to the Sun's Longitude + 6 Signs. The Heliocentric Longitudes and Latitudes for the other two Observations must be found in the same manner.

§ 30. Let the two Arches of Latitude ϖ, and ϖ'; be prolonged till they join in the Pole of the Ecliptic : the Angle formed by them at the Pole, will be equal to the motion of the Comet in Longitude, between the first and second Observation ; or $\varsigma' - \varsigma$. Then in the Spherical Triangle, whose Sides are the two Co-Latitudes, and the Angle between those Sides is given ; $= \varsigma' - \varsigma$; the third Side will be the angular motion of the Comet at the Sun, between the Radii Vectores, r, and r'. This Side is found by the common rules of Spherical Trigonometry, or the following Formula may be used.

$$\textit{Cof. } (\varsigma' - \varsigma). \textit{ Cof. } \varpi. \textit{ Cof. } \varpi' + \textit{Sin. } \varpi. \textit{ Sin. } \varpi. = \textit{Cof. } V.$$

V being the Angle at the Sun sought.

Let V', be the Angle at the Sun, formed by the first and third Radius Vector, r, and r'', then the same Formula will

is also constant, and therefore may be prepared apart, by adding the Logarithms of those two Quantities.

In the Triangle, No. 4, the Part of the Formula $\dfrac{\text{Sin. } \theta}{\text{Sin. S T C}}$ is constant, and may therefore be prepared apart, by subtracting the Log. Sin. of S T C from the Log. Sin. of θ.

give V, with the neceſſary changes of $\mathcal{C}''$ for $\mathcal{C}'$, and ϖ'' for ϖ. Thus.

$$Coſ. (\mathcal{C}'' - \mathcal{C}'). Coſ. \varpi. Coſ. \varpi'' + Sin. \varpi. Sin. \varpi'' = Coſ. V'.$$

§ 31. Now if the Perihelion Diſtance, and the Time of Arrival at the Perihelion, determined before, were exact ; then V, would be $= U$; and $V' = U'$. But as that can ſcarce ever be the caſe, let $U - V$ be called m, and $U' - V'$, be n.

§ 32. A ſecond Hypotheſis muſt therefore be formed, in which the time of the paſſage of the Comet by the Perihelion being unaltered, the Perihelion Diſtance muſt be changed a little ; a fiftieth for example. In this new Hypotheſis the Anomalies, v, v', and v'', and Radii Vectores, r, r', and r'', with the Angles between them, U, and U', muſt be computed as before ; and in the ſame manner with theſe new Radii Vectores ; V and V' muſt be found from the three Obſervations ; and let $U - V$ in this Hypotheſis be $= m'$, and let $U' - V'$ be $= n'$.

§ 32. A third Hypotheſis muſt now be formed, in which the Perihelion Diſtance of the firſt Hypotheſis muſt be reſtored ; but the time of the Paſſage at the Perihelion muſt be changed a little ; ſuppoſe half a day, or a day, more or leſs, according to the quantity of the errors. U, and U', and V, and V', muſt be anew computed as before, and $U - V$ muſt be called $= m''$; and $U' - V'$, muſt be called $= n''$.

§ 33. From the errors found in theſe three Hypotheſes ; the true Perihelion Diſtance, and true time of the Paſſage by the Perihelion, will be found as follows.

Let u, be the number by which the change in the Perihelion Diftance of the fecond Hypothefis, muft be multiplied to obtain the true correction of the Diftance; and t, the number by which the change of time in the Paffage by the Perihelion muft be multiplied to obtain the correction of the time of the Paffage by the Perihelion. The following Equations will give them.

$$u\ (m - m') + t\ (m - m'') = m.$$
$$u\ (n - n') + t\ (n - n'') = n.$$

From thefe Equations we obtain the following.

$$u = \frac{m\ (n - n'') - n\ (m - m'')}{(m - m')\ (n - n'') - (m - m'')\ (n - n')}$$

$$t = \frac{m - u\ (m - m')}{m - m''}.$$

Whence the true Perihelion Diftance and Time of Arrival at the Perihelion are eafily obtained.

§ 34. Thefe two Elements, which give the dimenfions of the Orbit, and real Motion of the Comet with refpect to the Sun, being afcertained; the others, which relate to the Pofition of the Orbit, with refpect to that of the Earth; are thus found. Let N, be the Place of the Afcending Node; and I, the Inclination of the Orbit; the following Equations will give them.

$$Tang.\ N = \frac{Tang.\ \varpi'\ Sin.\ \mathcal{E} - Tang.\ \varpi\ Sin.\ \mathcal{E}'}{Tang.\ \varpi'\ Cof.\ \mathcal{E} - Tang.\ \varpi\ Cof.\ \mathcal{E}'}$$

$$Tang.\ I = \frac{Tang.\ \varpi'}{Sin.\ (\mathcal{E}' - N)}$$

Or the following.

$$Tang.\ N = \frac{Tang.\ \varpi''\ Sin.\ \varepsilon - Tang.\ \varpi\ Sin.\ \varepsilon''}{Tang.\ \varpi''\ Cof.\ \varepsilon - Tang.\ \varpi\ Cof.\ \varepsilon''}$$

$$Tang.\ I = \frac{Tang.\ \varpi''}{Sin.\ (\varepsilon'' - N)}$$

If to determine the Angles N and I, the two laſt Formulæ be uſed; it is evident that the Tangent of N, may belong to the Angle N, or to the Angle $180° + N$; but the bare inſpection of the Figure by which the firſt Approximation was obtained, will ſhew at once which Angle ſhould be uſed.

§ 35. The Hypothenuſe of a Spherical right-angled Triangle, whoſe Baſe is $\varepsilon'' - N$, and Perpendicular ϖ''; will be the diſtance of the Comet from the Node at the third Obſervation; and the difference between this Hypothenuſe and v'', is the interval between the Node and Perihelion, counted on the Orbit; whence the Longitude of the Perihelion on the Orbit, is eaſily found.

§ 36. Mr. Pingrè obſerves, that when the firſt Approximation has given the Perihelion diſtance, and time of arrival at the Perihelion, not very near the truth, which will often be the caſe, if the arch of the Comet's Motion was too great, or the Obſervations uſed, not very accurate; the firſt three Hypotheſes will not give the Orbit of the Comet, perfectly correct; but differences will be ſtill found between the U, and V, in the fourth Hypotheſis formed with the corrected Perihelion diſtance, and time of Arrival. When this is the caſe, if abſolute preciſion is deſired, this fourth muſt be aſſumed as a firſt Hypotheſis, and two others muſt be anew computed with very ſmall variations in the Perihelion Diſtance, and time of Arrival. The reſult of this proceſs will give the Elements of

Q q

the Orbit of the Comet, to as great a degree of Exactnefs as the Obfervations made ufe of, will allow.

§ 37. The method of correction above given, fails in one cafe, viz. when the Radius Vector of the Comet is perpendicular, or very nearly fo, to the vifual Ray from the Earth to the Comet, T C. In this cafe, which though very rare, will fometimes happen it did happen, in the correction of the Elements of the Comet of 1790, Five Hypothefes muft be ufed. The Firft, with the Approximate Elements; the Second, with a very fmall variation of the Perihelion Diftance; the Third, with double that Variation; the Fourth, with a very fmall Variation in the time of the Comet's arrival at the Perihelion; the Fifth, with double that Variation.

Then let m, m', m'', m''', m'''' be the values of $U - V$, and n, n', n'', n''', n'''', be the Values of $U' - V'$, in the five Hypothefes; to determine the Values of u, and t; the following Equations muft be formed.

$$(4\,m' - 3\,m - m'')\,u + (m'' - 2\,m' + m)\,u^2 + (4\,m''' - 3\,m - m'''')\,t + (m'''' - 2\,m''' + m)\,t^2 + 2\,m = 0.$$

$$(4\,n' - 3\,n - n'')\,u + (n'' - 2\,n' + n)\,u^2 + (4\,n''' - 3\,n - n'''')\,t + (n'''' - 2\,n''' + n)\,t^2 + 2\,n = 0.$$

As however the refolution of thefe Equations, is very troublefome and tedious; it will be by much the beft way, when in the Calculation of the firft Hypothefis, any one of the Obfervations chofen, is found to be in the cafe above mentioned, to lay that Obfervation afide, and take another within a day or two of it, which will probably not be fo circumftanced, and proceed with the correction by three Hypothefes, as before mentioned.

§ 38. As the Letters ufed by Mr. De La Place, in his Method for the Orbits of Comets, differ from thofe ufed by Mr. Pingrè, it may be convenient to bring into one View the Letters ufed by both, in the whole Operation, with their Significations.

De la Place.	Pingrè.	In the Method for the obtaining the Approximate Perihelion Diſtance and Time of Arrival at the Perihelion.		
$'$	$''$	Mark of the Firſt	Obfervation, or Interval, or any other Quantity deduced from the Obferva- tions, throughout the Operations.	
$''$	$'''$	Second - -		
$'''$	$''''$	Third - -		
		Fourth - -		
$ε$	C	Geocentric Longitude	of the Comet.	
$γ$	L	Geocentric Latitude		
$τ$	$θ$	Time elapfed between each Obfervation.		
i	$τ$	Time elapfed between each Obfervation and the Epoch.		
$δ$	d	Quotient of the Difference between the Longitudes, or Latitudes, of the Comet; divided by the Elapfed Time.		
z	$ζ$	A Longitude or Latitude of the Comet, for a Moment diſtant from the Epoch by a few Days.		
a	$β$	The four Quantities formed from the Coefficients of z, and z^2, ufed in the Equations following.		
b	$γ$			
h	$μ$			
l	$ν$			
$ϰ$	C'	Geocentric Longitude of the Comet	At the Epoch.	
$θ$	L	Geocentric Latitude of the Comet		
$Λ$	T	Longitude of the Earth -		
R	R	Radius Vector of the Earth		

De La Place.	Pingré.	
R^1	R'	Radius Vector of the Earth for a Longitude, 90 Degrees from the Longitude at the Epoch.
x	Δ	The Assumed Curtate Distance of the Comet from the Earth at the Epoch.
r	r	Radius Vector of the Comet.
y	y	A Quantity found by the Second Equation.
P	B	A Quantity found by the first Equation for the Perihelion Distance of the Comet.
D	π	The Perihelion Distance of the Comet.
v	υ	The Anomaly of the Comet.

In the Correction of the Approximate Elements by Distant Observations.

De La Place.	Pingré.	
α	C	The Geocentric Longitude ⎫
θ	L	The Geocentric Latitude - ⎪
$\mathfrak{E}$	C	The Heliocentric Longitude ⎬ of the Comet.
ϖ	λ	The Heliocentric Latitude ⎪
ε	ζ	The Elongation - - ⎭
U	φ	Angle between the First and Second ⎫ Radius Vector.
U'	x	Angle between the First and Third ⎭
V	Φ	Motion of the Comet in its Orbit between the First and Second Observation.
V'	X	The same, between the Second and Third Observation.

De La Place.	Pingré.	
u	*x*	Number by which the Change of Perihelion Diſtance - - ⎫ Is to be multiplied to obtain the true Correction.
t	*y*	Number by which the Change of the Time of Arrival at the Perihelion ⎬
♅	*u*	Diſtance of the Comet from the Aſcending Node at any Obſervation ; meaſured on the Orbit.

In the Figure relative to the Correction.

S	*S*	The Sun.
T	*T*	The Place of the Earth.
C′	*K*	The Place of the Comet in its Orbit.
P	*C′*	The Place of the Comet projected on the Plane of the Ecliptic.

Application of Mr. De la Place's Method of finding the
Approximate Perihelion Diſtance, and time of Arrival
at the Perihelion, to the Comet of 1769.

§ 1. THE Obſervations made uſe of in this example, are
thoſe uſed in Mr. *Pingrè's Comêtographie*, as an example to Mr.
De la Place's method. It ſhould be here premiſed, that theſe
Obſervations being made at abſolutely equal Intervals, the
Coefficients of z, and z^2, are found by the Equation of § 12,
in the preceding Chapter; and all the calculation of the $\delta \xi$, and
$\delta \gamma$, &c; is ſuperfluous. In order however to make the Example as complete as poſſible, the coefficients of z, and z^2, are
here computed by both the methods.

§ 2. Let then the Obſervations choſen be the following five,
all made in September at 14 $\overset{H}{}$

Days.	Longitude of Comet.			Latitude of Comet.			Longitude of Sun.			Logarithms of Sun's Diſtance.
	°	′	″	°	′	″	°	′	″	
4	80	56	11	− 17	51	39	162	42	5	0 . 0031324
6	90	18	18	− 20	7	59	164	38	45	0 . 0029008
8	101	0	54	− 22	5	2	166	35	31	0 . 0026648
10	112	38	35	− 23	20	52	168	32	22	0 . 0024242
12	124	19	22	− 23	43	55	170	29	20	0 . 0021838

§ 3. We now proceed to the Calculation of the $\delta \xi$ as
follows,

$$Divisors$$

For the $\delta\mathfrak{C}$ $2 \cdot 2 \cdot 2 \cdot 2$
For the $\delta^2\mathfrak{C}$ $4 \cdot 4 \cdot 4$
For the $\delta^3\mathfrak{C}$ $6 \cdot 6$
For the $\delta^4\mathfrak{C}$ 8

$$
\begin{array}{ll}
\mathfrak{C} & \overset{\circ}{8}0 \cdot \overset{\prime}{5}6 \cdot \overset{\prime\prime}{1}1 \\
\mathfrak{C}' & 90 \cdot 18 \cdot 18 \\
\mathfrak{C}'' & 101 \cdot 0 \cdot 54 \\
\mathfrak{C}''' & 112 \cdot 38 \cdot 35 \\
\mathfrak{C}'''' & 124 \cdot 19 \cdot 22
\end{array}
$$

$$
\left.
\begin{array}{l}
\mathfrak{C}' - \mathfrak{C} = + \overset{\circ}{9} \cdot \overset{\prime}{2}2 \cdot \overset{\prime\prime}{7} \\
\mathfrak{C}'' - \mathfrak{C}' = + 10 \cdot 42 \cdot 36 \\
\mathfrak{C}''' - \mathfrak{C}'' = + 11 \cdot 37 \cdot 41 \\
\mathfrak{C}'''' - \mathfrak{C}''' = + 11 \cdot 40 \cdot 47
\end{array}
\right\}
\begin{array}{c} \text{dividing} \\ \text{by 2} \\ \text{we have} \end{array}
\left\{
\begin{array}{l}
\delta\mathfrak{C} = + \overset{\circ}{4} \cdot \overset{\prime}{4}1 \cdot \overset{\prime\prime}{3}, 5 \\
\delta\mathfrak{C}' = + 5 \cdot 21 \cdot 18, 0 \\
\delta\mathfrak{C}'' = + 5 \cdot 48 \cdot 50, 5 \\
\delta\mathfrak{C}''' = + 5 \cdot 50 \cdot 23, 5
\end{array}
\right.
$$

The Divisor 2, being the Interval between the First and Second, Second and Third, Third and Fourth, and Fourth and Fifth Observations.

§ 4. For the Computation of the $\delta^2\mathfrak{C}$, we have,

$$
\left.
\begin{array}{l}
\delta\mathfrak{C}' - \delta\mathfrak{C} = + \overset{\prime}{4}0 \cdot \overset{\prime\prime}{1}4, 5 \\
\delta\mathfrak{C}'' - \delta\mathfrak{C}' = + 27 \cdot 32, 5 \\
\delta\mathfrak{C}''' - \delta\mathfrak{C}'' = + 1 \cdot 33, 0
\end{array}
\right\}
\begin{array}{c} \text{dividing} \\ \text{by 4} \\ \text{we have} \end{array}
\left\{
\begin{array}{l}
\delta^2\mathfrak{C} = + \overset{\prime}{1}0 \cdot \overset{\prime\prime}{3}, 625 \\
\delta^2\mathfrak{C}' = + 6 \cdot 53, 125 \\
\delta^2\mathfrak{C}'' = + 0 \cdot 23, 25
\end{array}
\right.
$$

The Divisor 4, being the Interval between the Third and First, Fourth and Second, and Fifth and Third Observations.

§ 5. For the Computation of the $\delta^3\mathfrak{C}$, we have,

$$\delta^2\mathbb{C}' - \delta^2\mathbb{C} = -\overset{'}{3} . \overset{''}{10}, 5$$
$$\delta^2\mathbb{C}'' - \delta^2\mathbb{C}' = -\overset{'}{6} . 29, 875$$
$$\left.\begin{array}{l}\end{array}\right\}\begin{array}{c}\text{dividing}\\ \text{by 6}\\ \text{we have}\end{array}\left\{\begin{array}{l} \delta^3\mathbb{C} = -\overset{'}{0} . \overset{''}{31}, 8166 \&\text{c.}\\ \delta^3\mathbb{C}' = -1 . \overset{''}{4}, 9791 \end{array}\right.$$

The Divifor 6, being the Interval between the Fourth and Firft, and Fifth and Second Obfervations.

§ 6. For the Computation of $\delta^4\mathbb{C}$, we have,

$$\delta^3\mathbb{C}' - \delta^3\mathbb{C} = -\overset{'}{0} . \overset{''}{33} . 1625 \begin{bmatrix}\text{dividing}\\ \text{by 8}\\ \text{we have}\end{bmatrix} \delta^4\mathbb{C} = -\overset{'}{0} . \overset{''}{4}, 1453$$

The Divifor 8, being the Interval between the Fifth and Firft Obfervations.

§ 7. For the formation of the Coefficients of z, and z^2, let the Epoch be fixed at the third Obfervation; then according to § 7, in the preceding Chapter, we fhall have

$$\left.\begin{array}{l} i = -4\\ i' = -2\\ i'' = 0\\ i''' = +2\\ i'''' = +4 \end{array}\right\} \text{ and } \left\{\begin{array}{l} i + i' = -6; \ i + i' + i'' = -6\\ i\, i' = +8; \ i\, i'''' = -8; \ i'\, i''' = -4\\ i\, i'\, i''' = +16. \end{array}\right.$$

§ 8. Now for the coefficients of z, in the formula

$$z\left\{\delta\mathbb{C} - (i + i')\, \delta^2\mathbb{C} + (i\, i')\, \delta^3\mathbb{C} - (i\, i'\, i''')\, \delta^4\mathbb{C}\ ;\ \text{we have,}\right.$$

$$\delta\mathbb{C} = +\overset{\circ}{4} . \overset{'}{41} . \overset{''}{3}, 5 \quad . \quad \delta^2\mathbb{C} = +\overset{'}{10} . \overset{''}{3}, 625$$
$$i + i' = - \qquad\qquad 6$$
$$\overline{\qquad -60 . 21 . 750\qquad}$$

$$\delta^3 6 = - \overset{'}{0} . \overset{''}{3} 1, \ 8166$$
$$i \, i' = + \qquad\qquad 8$$
$$\overline{ - 4 . 14 . 5328}$$

$$\delta^4 6 = - \overset{'}{0} . \overset{''}{4}, \ 1453$$
$$i \, i' \, i''' = + \qquad\qquad 16$$
$$\overline{ - 1 . 6, \ 3248}$$

And reducing all the Quantities to Seconds, and prefixing the proper Signs, the formula becomes

$$z \{ + \overset{''}{16}863, 5 + \overset{''}{3}621, 750 - \overset{''}{2}54, 5328 + \overset{''}{6}6, 3248 \} = + \overset{''}{2}0297$$

§ 9. In like manner for the Coefficients of z^2 in the formula,

$$z^2 \left\{ \delta^2 6 - (i + i') \, \delta^3 6 + (i \, i' + i \, i'' + i' \, i''') \, \delta^4 6 \right\} \text{ we have, }\cdot$$

$$\delta^2 6 = + \overset{'}{10} . \overset{''}{3}, \ 625$$

$$\delta^3 6 = - \overset{'}{0} . \overset{''}{3} 1, \ 8166$$
$$i + i' = - \qquad\qquad 6$$
$$\overline{ + 3 . 10, \ 8996}$$

$$\delta^4 6 =$$
$$i \, i + i \, i''' + i' \, i''' = - \qquad \dfrac{- \ 0 . \ \overset{''}{4}, \ 1453}{4}$$
$$\overline{ + 0 . 16 . 5812}$$

And reducing all the Quantities to Seconds, and prefixing the proper Signs; the formula becomes,

$$z^2 \{ + \overset{''}{6}03, \ 625 - \overset{''}{1}90, \ 8996 + \overset{''}{1}6 . 5812 \} = + \overset{''}{4}29, \ 3.$$

§ 10. The whole formula then of § 11 in the laſt Chapter becomes

$$\overset{\circ}{101} . \overset{'}{0} . \overset{''}{5}4 + z \ 20\overset{''}{2}97 . + z^2 \ \overset{''}{4}29, \ 3 .$$

S ſ

§ 11. It may not be improper here to obferve that had the Longitudes of the Comet been decreafing, the $c'-c$ &c. would have been Negative. In all cafes ftrict regard muft be had, to the proper application of the Signs.

§ 12. We fhall now, in order to make the Example complete, compute the Coefficients of z and z^2 by the formula

$$c'' + z\left(\frac{c - c'''}{12\,i} + \frac{2\,c''' - 2\,c'}{3\,i}\right) + z^2\left(\frac{2\,c' + 2\,c'''}{3\,i^2} - \frac{5\,c''}{4\,i^2} - \frac{c''' + c}{24\,i^2}\right)$$

§ 13. For the Coefficients of z we have

	$\overset{\circ}{}$	$\overset{\prime}{}$	$\overset{\prime\prime}{}$			$\overset{\circ}{}$	$\overset{\prime}{}$	$\overset{\prime\prime}{}$
c	80	56	11		$2\,c'''$	225	17	10
c'''	124	19	22		$2\,c'$	180	36	36
$-$	43	23	11		$+$	44	40	34

$12\,i = + 24\,(\,\overset{\prime\prime}{156191}\,)$

$\dfrac{c - c'''}{12\,i} = -6508$

$2\,c''' - 2\,c' \quad 3\,i = 6\,)\,160834$

$\dfrac{2\,c''' - 2\,c'}{3\,i} = + 26805,\,66$

The Formula then becomes

$$z\left\{-\overset{\prime\prime}{6}508 + 26805,\,66\right\} = + \overset{\prime\prime}{20}297,\,7$$

§ 14. For the Coefficients of z^2 we have,

	$\overset{\circ}{}$	$\overset{\prime}{}$	$\overset{\prime\prime}{}$			
$2\,c'$	180	36	36		$5\,c''$	505 . 4 . 30
$2\,c'''$	225	17	10		$4\,i^2 =$	16) $1818\overset{\prime\prime}{2}70$
	405	53	46		$\dfrac{5\,c''}{4\,i^2} =$	113641, 875

$\dfrac{2\,c' + 2\,c'''}{3\,i^2} \quad 3\,i^2 = 12\,(\,1461\overset{\prime\prime}{2}26$

$= + 121768,\,833$

c''''		124 . 19 . 22
c		80 . 56 . 11
		205 . 15 . 33

$24\,i^2 = 96)\,738933$

$\dfrac{c''' + c}{24\,i^2} = 7697,\,22$

The formula then becomes,

$$z^2 \left\{ + 121\overset{''}{7}68, 833 - 113\overset{''}{6}41, 875 - 7697'', 22 \right\} = + 42\overset{''}{9}, 8$$

§ 15. The formula for the Longitudes is therefore by this method,

$$10\overset{o}{1} \cdot \overset{'}{0} \cdot \overset{''}{5}4 + z\, 20\overset{''}{2}97, 7 + z^2\, 42\overset{''}{9}, 8.$$

The same to the Fraction of a Second as by the other method.

§ 16. The same operation must now be performed on the Latitudes. And first for the δ γ we have as follow :

Divisors

For the $\delta\gamma$	2 . 2 . 2 . 2
For the $\delta^2\gamma$	4 . 4 . 4
For the $\delta^3\gamma$	6 . 6
For the $\delta^4\gamma$	8

	o	'	''
γ	17	51	39
γ'	20	7	59
γ''	22	5	2
γ'''	23	20	52
γ''''	23	43	55

$$
\left.
\begin{array}{l}
\gamma' - \gamma = - 2 \cdot 16 \cdot 20 \\
\gamma'' - \gamma' = - 1 \cdot 57 \cdot 3 \\
\gamma''' - \gamma'' = - 1 \cdot 15 \cdot 50 \\
\gamma'''' - \gamma''' = - 0 \cdot 23 \cdot 3
\end{array}
\right\}
\begin{array}{c}
\text{dividing} \\
\text{by 2} \\
\text{we have}
\end{array}
\left\{
\begin{array}{l}
\delta\gamma = - 1 \cdot 8 \cdot 10 \\
\delta\gamma' = - 0 \cdot 58 \cdot 31, 5 \\
\delta\gamma'' = - 0 \cdot 37 \cdot 55 \\
\delta\gamma''' = - 0 \cdot 11 \cdot 31, 5
\end{array}
\right.
$$

§ 17. For the $\delta^2\gamma$, we have,

$$
\left.
\begin{array}{l}
\delta\gamma' - \delta\gamma = + \overset{'}{9}.\overset{''}{38},5 \\
\delta\gamma'' - \delta\gamma' = +20.36,5 \\
\delta\gamma''' - \delta\gamma'' = +26.23,5
\end{array}
\right\}
\begin{array}{c}
\text{dividing} \\
\text{by 4} \\
\text{we have}
\end{array}
\left\{
\begin{array}{l}
\delta^2\gamma = + \overset{'}{2}.\overset{''}{24},625 \\
\delta^2\gamma' = + 5.9,125 \\
\delta^2\gamma'' = + 6.35,875
\end{array}
\right.
$$

§ 18. For the $\delta^3\gamma$, we have,

$$
\left.
\begin{array}{l}
\delta^2\gamma' - \delta^2\gamma = + 2.\overset{''}{44},500 \\
\delta^2\gamma'' - \delta^2\gamma' = + 1.26,750
\end{array}
\right\}
\begin{array}{c}
\text{dividing} \\
\text{by 6} \\
\text{we have}
\end{array}
\left\{
\begin{array}{l}
\delta^3\gamma = + \overset{'}{0}.\overset{''}{27},4166 \\
\delta^3\gamma' = + 0.14,4583
\end{array}
\right.
$$

§ 19. For the $\delta^4\gamma$, we have,

$$
\delta^3\gamma' - \delta^3\gamma = - \overset{''}{12}.9583
\quad
\left\{
\begin{array}{c}
\text{dividing} \\
\text{by 8} \\
\text{we have}
\end{array}
\right\}
\quad
\delta^4\gamma = - \overset{''}{1},6198
$$

§ 20. The Values of i and its multiples, being the same as in the former procefs; for the Coefficients of z in the formula,

$$
z\left\{\delta\gamma - (i + i')\,\delta^2\gamma + (i\,i')\,\delta^3\gamma - (i\,i'\,i''')\,\delta^4\gamma\right\}; \text{ we have,}
$$

$$
\delta\gamma = - 1.8.10, \qquad
\begin{array}{r}
\delta^2\gamma = + \overset{'}{2}.\overset{''}{24},625 \\
i + i' = - 6 \\
\hline
- 14.27,750
\end{array}
$$

$$
\begin{array}{r}
\delta^3\gamma = + \overset{'}{0}.\overset{''}{27},4166 \\
i\,i' = + 8 \\
\hline
+ 3.39,3328
\end{array}
\qquad
\begin{array}{r}
\delta^4\gamma = - \overset{'}{.0}.\overset{''}{1},6198 \\
i\,i'\,i''' = + 16 \\
\hline
- 0.25,9168
\end{array}
$$

And reducing all the Quantities to Seconds, and prefixing the proper Signs, the formula becomes

$$z\{-40\overset{..}{9}0+8\overset{..}{6}7,\ 750+2\overset{.}{1}9,\ 3328+2\overset{..}{5},\ 9168\}=-2\overset{..}{9}77$$

§ 21. In like manner for the Coefficients of z^2 in the formula,

$$z^2\left\{\delta^2\gamma-(i+i')\,\delta^3\gamma+(i\,i'+i\,i'''+i'\,i''')\,\delta^4\gamma\right\}\text{ we have,}$$

$$\delta^2\gamma=+\overset{.}{2}\ .\ 2\overset{..}{4},\ 625 \qquad\qquad \delta^3\gamma=+\overset{.}{0}\ .\ 2\overset{..}{7},\ 4166$$
$$i+i'=-6$$
$$\overline{-2\ .\ 44,\ 4996}$$

$$\delta^4\gamma=$$
$$i\,i'+i\,i'''+i'\,i'''=-\overset{..}{0}\ .\ \overset{..}{1},\ 6198$$
$$4$$
$$\overline{+\ 0\ .\ \ 6,\ 4792}$$

And reducing all the Quantities to Seconds, and prefixing the proper Signs; the formula becomes

$$z^2\left\{+1\overset{..}{4}4,\ 625+1\overset{..}{6}4,\ 4996+\overset{..}{6},\ 4792\right\}=+3\overset{..}{1}5,\ 6.$$

§ 22. The whole formula therefore for the Latitudes according to § 11, and § 13 of the last Chapter becomes

$$2\overset{o}{2}\ .\ \overset{.}{5}\ .\ \overset{..}{2}-z\ 29\overset{..}{7}7\ .\ +z^2\ 3\overset{..}{1}5,\ 6.$$

The Computation of these quantities by the formula of § 12 in the last Chapter, is so easy, as not to require a repetition of the process as given at large for the Longitudes in § 12, &c.

§ 23. As five observations have been used, and the Epoch is fixed at the third Observation; and as the Longitudes and

T t

Latitudes of the Comet, have been made ufe of inftead of its right Afcenfions and Declinations; the only ufe of the Coefficients of z and z^2 is to form from them the Quantities a, b, b, and l, which we fhall proceed to do, according to the directions given in § 15, of the laft Chapter.

In the formula for the Longitudes we have

Coefficient of z 20297, 7	Log.		4 . 3074470
Conftant Log.	Subtract		3 . 5500081
a Log.		+	0 . 7574389

Though Mr. Pingrè before calls the Coefficient of z^2 429, 8 yet in forming the Quantity b, he ufes it 429, 75, which really is rather more exact; we fhall therefore follow him in this, and fay,

Twice the Coefficient of z^2 859, 5	Log.		2 . 9342459
Conftant Log.	Subtract		1 . 7855911
b Log.		+	1 . 1486548

In the formula for the Latitudes, we have,

Coefficient of z — 2977	Log.		3 . 4737788
Conftant Log.	Subtract		3 . 5500081
b Log.		—	9 . 9237707

Twice the Coefficient of z^2 631, 2	Log		2 . 8001670
Conftant Log.	Subtract		1 . 7855911
l Log.		+	1 . 0145759

§ 24. We shall now collect the Quantities necessary for the Solution of the following Equations according to § 16 in the preceding Chapter.

Longitude of the Comet at the Epoch $\qquad = \alpha \qquad$ 101 . 0 . 54''

Latitude of the Comet at the Epoch $\qquad = \theta \quad -$ 22 . 5 . 2

Longitude of Earth at the Epoch $\qquad = A \qquad$ 346 . 35 . 31

Log. of Radius Vector of the Earth at the Epoch $= R \quad$ 0, 0026636

Log. of R. V. of Earth 90° beyond the Epoch $= R' \quad$ 9 . 9931810

$$\begin{array}{lll} \text{Log.} & a + & 0 . 7574389 \\ \text{Log.} & b + & 1 . 1486548 \\ \text{Log.} & h - & 9 . 9237707 \\ \text{Log.} & l + & 1 . 0145759 \end{array}$$

§ 25. As in the different trials necessary for the solution of the following Equations, the Value of the Coefficients of the unknown Quantities remains the same; these Coefficients had better be first computed apart; as the Operation will thereby be both shortened and simplified.

Coefficients of the first Equation § 17. last Chapter.

	LOGARITHMS.	NUMBERS.
Cos θ - -	— 9 . 9669084	
Cos θ^2 - - -	+ 9 . 9338168	+ 0 . 858651
R - -	+ 0 . 0026636	
2 - -	+ 0 . 3010300	
Cos $(A - \alpha)$ -	— 9 . 6164449	
2 R. Cos $(A - \alpha)$	— 9 . 9201385	— 0 . 832029
R^2 - -	+ 0 . 0053272	+ 1 . 012342

And the Equation becomes

$$r^2 = + \frac{x^2}{0.858651} - 0.832029\,x + 1.012342$$

§ 26. The Coefficients of the Second Equation; are

			LOGARITHMS.	NUMBERS.
Sin $(A - \alpha)$	-		$-\,9.9592882$	
a	-	-	$+\,0.7574389$	
2	-	-	$+\,0.3010300$	
R^2	-	-	$+\,0.0053272$	
$2\,a\,R^2$	-		$+\,1.0637961$	
$\dfrac{\text{Sin}\,(A-\alpha)}{2\,a\,R^2}$	-		$-\,8.8954921$	$-\,0.0786126$
R	-	-	$+\,0.0026636$	
Sin $(A - \alpha)$	-		$-\,9.9592882$	
R Sin $(A - \alpha)$			$-\,9.9619518$	
$2\,a$	-	-	$+\,1.0584689$	
$\dfrac{R\,\text{Sin}\,(A-\alpha)}{2\,a}$	-		$-\,8.9034829$	$-\,0.0800724$
b	-	-	$+\,1.1486548$	
$2\,a$	-	-	$+\,1.0584689$	
$\dfrac{b}{2\,a}$	-	-	$+\,0.0901859$	$+\,1.230796$

And the Equation becomes

$$y = -0.0786126 + \frac{0.0800724}{r^3} - 1.230796\,x$$

§ 27. The Coefficients of the Third Equation are

	LOGARITHMS.	NUMBERS.
a^2 - -	$+ 1 . 5148778$	$+ 32 . 724860$
Tang. θ - -	$- 9 . 6082375$	$- 0 . 4057304$
h - -	$- 9 . 9237707$	
Cos θ² - -	$+ 9 . 9338168$	
$\dfrac{\text{`} h}{\text{Cos } θ^2}$ - -	$- 9 . 9899539$	$- 0 . 9771335$
$R' - 1$ -	$- 8 . 1925295$	
Cos $(A - \alpha)$ - -	$- 9 . 6164449$	
$(R' - 1)$ Cos $(A - \alpha)$	$+ 7 . 8089744$	$+ 0 . 00644131$
Sin $(A - \alpha)$ -	$- 9 . 9592882$	
R - -	$+ 0 . 0026636$	
$\dfrac{\text{Sin } (A - \alpha)}{R}$ -	$- 9 . 9566256$	$- 0 . 9049520$
$(R' - 1)$Cos$(A - \alpha) - \dfrac{\text{Sin}(A - \alpha)}{R}$		$+ 0 . 9113933$
$R' - 1$ - -	$- 8 . 1925295$	
Sin $(A - \alpha)$ -	$- 9 . 9592882$	
$(R' - 1)$ Sin $(A - \alpha)$	$+ 8 . 1518177$	$+ 0 . 0141846$
Cos $(A - \alpha)$ -	$- 9 . 6164449$	
R - -	$+ 0 . 0026636$	
$\dfrac{\text{Cos}(A - \alpha)}{R}$ - -	$- 9 . 6137813$	$- 0 . 4109428$
$(R' - 1)$Sin$(A - \alpha) + \dfrac{\text{Cos } (A - \alpha)}{R}$	$- 9 . 5985259$	$- 0 . 3967582$
$2 a$ - -	$+ 1 . 0584689$	
$2a \left\{ (R' - 1)\text{Sin}(A - \alpha) + \text{Cos}\dfrac{(A - \alpha)}{R} \right\}$	$- 0 . 6569948$	$- 4 . 539361$
R^2 - -	$+ 0 . 0053272$	
$\dfrac{1}{R^2}$ - -	$+ 9 . 9946728$	$+ 0 . 9878087$

U u

And the Equation becomes

$$0 = y^2 + 32.72486 \, x^2 + (-0.4057304 \, y - 0.9771335 \, x)^2$$

$$+ 1.8227866 \, y - 4.539361 \, x + 0.9878087 - \frac{2}{r}$$

§ 28. Though, as the Fourth Equation in the prefent Example ferves only to verify the refult of the others, as explained § 19 of the preceding Chapter, it is not neceffary to calculate the Coefficients apart; ufe being made of that Equation only once; yet it feems as well to give that computation in the fame manner as the others; it being ufed to find y, in cafes where the Motion of the Comet in Latitude, much exceeds its Motion in Longitude; in which cafe Equation 2 ferves as a Verification of the Procefs.

The Coefficients then of the Fourth Equation are

	LOGARITHMS.	NUMBERS.
b - -	$- 9.923707$	
Tang. θ - -	$- 9.6082375$	
b Tang. θ - -	$+ 9.5320082$	$+ 0.3404146$
l - -	$+ 1.0145759$	
2 - -	$+ 0.3010300$	
b - -	$- 9.923707$	
$2 b$ - -	$- 0.2248007$	
$\dfrac{l}{2b}$ - -	$- 0.7897752$	$- 6.16276$

	LOGARITHMS.	NUMBERS.
a^2 - -	$+ 1 . 5148778$	
Sin θ - -	$- 9 . 5751460$	
Cos θ - -	$- 9 . 9669084$	
a^2 Sin θ. Cos θ. -	$- 1 . 0569322$	
$2 h$ - -	$- 0 . 2248007$	
$\dfrac{a^2 \text{ Sin θ. Cos θ.}}{2 h}$ -	$+ 0 . 8321315$	$+ 6 . 794093$
		$+ 0 . 340415$
		$+ 7 . 134508$
		$- 6 . 162760$
$h \text{ Tang. } \theta + \dfrac{l}{2h} + \dfrac{a^2 \text{ Sin θ Cos θ}}{2h}$	$+ 9 . 9875657$	$+ 0 . 971748$
R - -	$+ 0 . 0026636$	
Sin θ - -	$- 9 . 5751460$	
Cos θ - -	$- 9 . 9669084$	
R Sin θ. Cos θ. -	$- 9 . 5447180$	
$2 h$ - -	$- 0 . 2248007$	
$\dfrac{R \text{ Sin θ Cos θ}}{2 h}$ - -	$+ 9 . 3199173$	
Cos $(A-\alpha)$ -	$- 9 . 6164449$	
$\dfrac{R \text{ Sin θ. Cos θ.}}{2 h}$ Cos. $(A-\alpha)$	$- 8 . 9363622$	$- 0 . 086369$
R^3 - -	$+ 0 . 0079908$	
$\dfrac{1}{R^3}$ - -	$9 . 9920092$	$+ 0 . 981769$

And the Equation becomes

$$y = - 0 . 971748 \, x - 0 . 086369 \left(\frac{1}{r^3} - 0 . 981769 \right).$$

§ 29. Having now prepared the Coefficients of the Equations it remains to make a firſt ſuppoſition for x, the Curtate diſtance of the Comet from the Earth. Let the firſt Suppoſition be $x = $ 0, 3; the Solution of the Equations will give to Equation 3, a negative Remainder. Suppoſing $x = $ 0, 31; the Equation 3 will have a poſitive Remainder rather greater than the negative Remainder of the firſt Suppoſition. It is then evident that the true x, is between 0, 30 and 0 . 31 but nearer 0 . 30 than 0 . 31. Suppoſe then $x = $ 0, 3048, the Remainder will be Negative, but ſmall; and by proportional parts of the Change in the Suppoſition of x, and the Change of Remainder of Equation 3, we ſhall come to $x = $ 0, 3048303. As the operation in all the Suppoſitions, is preciſely the ſame; it was unneceſſary to give it at length more than once. We ſhall therefore apply the $x = $ 0 . 3048303 to the Solution of the Equations as follows.

§ 30. The firſt Equation is

$$r^2 = + \frac{x^2}{0 \cdot 85861} - 0 \cdot 832029\, x + 1 \cdot 012342.$$

	LOGARITHMS.	NUMBERS.
$+ x$ - -	$+ 9 \cdot 4840581$	
$+ x^2$ - -	$+ 8 \cdot 9681162$	
$+ 0 \cdot 85861$ -	$+ 9 \cdot 9338168$	
$\dfrac{x^2}{0 \cdot 85861}$ - -	$+ 9 \cdot 0342994$	$+ 0 \cdot 108218$
$- 0 \cdot 832029$ -	$- 9 \cdot 9201385$	
$+ x$ - -	$+ 9 \cdot 4840581$	
$0 \cdot 832029\, x$ -	$- 9 \cdot 4041966$	$- 0 \cdot 253628$
And collecting the Poſitive and Negative Quantities we have		$+ 1 \cdot 012342$
		$+ 0 \cdot 108218$
		$+ 1 \cdot 120560$
		$- 0 \cdot 253628$
r^2 - -	$+ 9 \cdot 9379850$	$+ 0\ \ 866932$
r - -	$+ 9 \cdot 9689925$	$+ 0 \cdot 9310919$

§ 31. The Second Equation is

$$y = -0 \cdot 0786126 + \frac{0 \cdot 0800724}{r^3} - 1 \cdot 230796 \; x$$

		LOGARITHMS.	NUMBERS.
$+$ 0 . 0800724	-	$+$ 8 . 9034829	
$+$ r^3	- -	$+$ 9 . 9069775	
$\dfrac{0 \cdot 0800724}{r^3}$	- -	$+$ 8 . 9965054	$+$ 0 . 0991986
$-$ 1 . 230796	-	$-$ 0 . 0901859	
$+$ x	- -	$+$ 9 . 4840581	
$-$ 1 . 230796 x		$-$ 9 . 5742440	$-$ 0 . 3751837
And collecting the Positive and Negative Quantities we have $\}$			$-$ 0 . 0786126
			$-$ 0 . 3751837
			$-$ 0 . 4537963
			$+$ 0 . 0991986
y *	- -	$-$ 9 . 5497355	$-$ 0 . 3545977

* Note. Mr. Pingrè has committed a most perplexing error (Vol. 2, Page 332) by giving the Value of y, as found by the Fourth Equation, for the Result of the Second Equation, and the same error is repeated a few lines lower, when he compares the Values of y, as found by the Two Equations. The Fourth Equation gives $y = -318$, &c. and the Second Equation, gives $y = -354$, &c.

X x

§ 32. The Third Equation is

$$0 = y^2 + 32 \cdot 72486\, x^2 + (- 0 \cdot 4057304\, y - 0 \cdot 9771335\, x)^2$$

$$+ 1 \cdot 8227866\, y - 4 \cdot 539361\, x + 0 \cdot 9878081 - \frac{2}{r}$$

	LOGARITHMS.	NUMBERS.
y^2 - -	$+ 9 \cdot 0994710$	$+ 0 \cdot 1257393$
$+ 32 \cdot 72486$ -	$+ 1 \cdot 5148778$	
x^2 - -	$+ 8 \cdot 9681162$	
$+ 32 \cdot 72486\, x^2$ -	$+ 0 \cdot 4829940$	$+ 3 \cdot 040843$
$- 0 \cdot 4057304$ -	$- 9 \cdot 6082375$	
$- y$ - -	$- 9 \cdot 5497355$	
$- 0 \cdot 4057304\, y$	$+ 9 \cdot 1579730$	$+ 0 \cdot 143871$
$- 0 \cdot 9771335$ -	$- 9 \cdot 9899539$	
$+ x$ - -	$+ 9 \cdot 4840581$	
$- 0 \cdot 9771335\, x$	$- 9 \cdot 4740120$	$- 0 \cdot 297860$
$- 0 \cdot 4057304\, y - 0 \cdot 9771335\, x)$	$- 9 \cdot 1874898$	$- 0 \cdot 153989$
$- 0 \cdot 4057304\, y - 0 \cdot 9771335\, x)^2$	$+ 8 \cdot 3749796$	$+ 0 \cdot 02371262$
$+ 1 \cdot 8227866$ -	$+ 0 \cdot 2607358$	
$- y$ - -	$- 9 \cdot 5497355$	
$1 \cdot 8227866\, y$ -	$- 9 \cdot 8104713$	$- 0 \cdot 6463551$
$- 4 \cdot 539361$ -	$- 0 \cdot 6569948$	
$+ x$ - -	$+ 9 \cdot 4840581$	
$4 \cdot 539361\, x$ -	$- 0 \cdot 1410529$	$- 1 \cdot 383735$
$- 2$ - -	$- 0 - 3010300$	
r - -	$+ 9 \cdot 9689925$	
$\frac{2}{r}$ - -	$+ 0 \cdot 3320375$	$- 2 \cdot 148016$

	NUMBERS. +	NUMBERS. −
	0 . 1257393	
	3 . 0408430	0 . 6463551
And collecting the Positive	0 . 0237126	1 . 3837350
and Negative Quantities	0 . 9878087	2 . 1480160
we have	4 . 1781036	4 . 1781061
Difference - -		— 0 . 0000025

This difference is so extremely small, that it may be reckoned as nothing. We therefore go on to the Fourth Equation as a proof of the truth of the former Operations.

§ 33. The Fourth Equation is

$$y = - 0 . 971748 \; x \; - \; 0 . 086369 \left(\frac{1}{r^3} - 0 . 981769 \right).$$

	LOGARITHMS.	NUMBERS.
— 0 . 971748 -	— 9 . 9875657	
+ x. - -	+ 9 . 4840581	
0 . 971748 x -	— 9 . 4716238	— 0 . 2962264
r^3 - -	+ 9 . 9069775	
$\frac{1}{r^3}$ - -	+ 0 . 0930225	+ 1 . 238860
— 0 . 981769 -		— 0 . 981769
	+ 9 . 4100869	+ 0 . 257091
— 0 . 086369 -	— 8 . 9363622	
0 . 086369$\left(\frac{1}{r^3} - 0 . 981769 \right)$	— 8 . 3464491	— 0 . 0222049
And collecting the Quantities,		— 0 . 2962264
we have		— 0 . 0222049
y - -		— 0 . 3184313

Which is not very different from the Value of *y* determined by Equation 2. Perhaps, fays Mr. Pingrè, the difference would have been ftill lefs, had not the Motion of the Comet been fo confiderable in Longitude, as $43,$ in the interval of time, included by the Five Obfervations.

§ 34. Therefore on the 8th of September, 1769, at 14 Hours Mean Time, on the Meridian of Paris; the Curtate Diftance of the Comet from the Earth, x, was 0 . 3048303; and the Radius Vector, *r*, was 0, 9310919.

§ 35. From the Radius Vector, Curtate Diftance and the Quantities determined in the foregoing Equations, we fhall now proceed to find the Perihelion Diftance of the Comet, and its Anomaly for the Epoch; whence the Time of the Paffage of the Comet by the Perihelion will be found. The Equations for this Purpofe, are as follow. (See § 21, and 22, of Mr. De La Place's Method.

EQUATION 1.

$$P = \frac{x}{Cof. \, \theta^2}\left(y + b \, x \, tang \, \theta \right) + R \, y \, Cof. \, (A - \alpha)$$

$$+ \, x \left\{ (R' - 1) \, Cof. \, (A - \alpha) - \frac{Sin \, (A - \alpha)}{R} \right\} + R \, a \, x \, Sin \, (A - \alpha)$$

$$+ \, R \, (R' - 1).$$

EQUATION 2.

$$D = r \quad \tfrac{1}{2} P^2$$

D being the Perihelion Diſtance of the Comet.

§ 36. For the Solution of the firſt Equation we have ;

	LOGARITHMS.	NUMBERS.
x - - -	$+ 9 . 4840581$	
$\text{Cos } \theta^2$ - -	$+ 9 . 9338168$	
$\dfrac{x}{\text{Cos } \theta^2}$ - -	$+ 9 . 5502413$	
y - -		$- 0 . 3545977$
b - -	$- 9 . 9237707$	
x - -	$+ 9 . 4840581$	
$\text{Tang } \theta$ - -	$- 9 . 6082375$	
$b \, x \, \text{Tang } \theta$ -	$+ 9 . 0160663$	$+ 0 . 1037687$
		$- 0 . 3545977$
$y + b \, x \, \text{Tang } \theta$ -	$- 9 . 3993778$	$- 0 . 2508290$
$\dfrac{x}{\text{Cos } \theta^2}$ - -	$+ 9 . 5502413$	
$\dfrac{x}{\text{Cos } \theta^2} \, (y + b \, x \, \text{Tang } \theta)$	$- 8 . 9496191$	$- 0 . 08904700$
R - -	$+ 0 . 0026636$	
y - -	$- 9 . 5497355$	
$\text{Cos } (A - \alpha)$ -	$- 9 . 6164449$	
$R \, y \, \text{Cos } (A - \alpha)$ -	$+ 9 . 1688440$	$+ 0 . 1475176$
x - -	$+ 9 . 4840581$	
$(R' - 1) \text{Cos} (A - \alpha) - \dfrac{\text{Sin } (A - \alpha)}{R}$ See the Coefficients of Equation (3) § 27	$+ 9 . 9597059$	
$x \left\{ (R' - 1) \text{Cos} (A - \alpha) - \dfrac{\text{Sin } (A - \alpha)}{R} \right\}$	$+ 9 . 4437640$	$+ 0 .. 2778203$

Y y

	LOGARITHMS.	NUMBERS.
R - -	$+$ 0 . 0026636	
a - -	$+$ 0 . 7574389	
χ - -	$+$ 9 . 4840581	
Sin $(A-u)$ - -	$-$ 9 . 9592882	
$R\ a\ \chi$ Sin $(A-u)$ -	$-$ 0 . 2034488	$-$ 1 . 5975291
R - -	$+$ 0 . 0026636	
$R'-1$ - -	$-$ 8 . 1925295	
$R\ (R'-1)$ - -	$-$ 8 . 1951931	$-$ 0 . 156745

And the Equation becomes

$$P = - \ 0 \ . \ 089047 + 0 \ . \ 1475176 + 0 \ . \ 2778203$$

$$- \ 1 \ . \ 5975291 - 0 \ . \ 0156745.$$

And collecting the Positive and Negative Quantities

$+$	$-$
0 . 1475176	0 . 08904700
0 . 2778203	1 . 59752910
0 . 4253379	0 . 01567450
	1 . 7022506
	0 . 4253379
	1 . 2769127

P - -

The Sign of P, in this Equation, being negative; the Comet has not yet passed the Perihelion.

§ 37. For the Solution of the Second Equation we have,

	LOGARITHMS.	NUMBERS.
P - -	— 0 . 1061613	
P^2 - -	+ 0 . 2123226	+ 1 . 630507
$\frac{1}{2} P^2$ - -		+ 0 . 8152535
r - - -		+ 0 . 9310919
D - -	+ 9 · 0638526	+ 0 . 1158384

§ 38. Having now the Perihelion Diftance of the Comet, $D = 0 . 1158384$, and the Radius Vector for the Epoch, $r = 0 . 9310919$, we fhall find the Anomaly, and Time of Arrival at the Perihelion, by the Rule of § 22 in the former Chapter.

For the Anomaly we have the Equation,

$$Cof. \; v = \frac{2 D}{r} - 1$$

	LOGARITHMS.	NUMBERS.
$2 D$ - -	+ 9 . 3648826	0 . 2316768
r - -	+ 9 . 9689925	
$\dfrac{2 D}{r}$ - -	+ 9 . 3958901	+ 0 . 2488228
$\dfrac{2 D}{r} - 1 = Cos\; v$ -	— 9 . 8757424	— 0 ..7511772
v - - -		$\overset{\circ}{138} . \overset{'}{41} . \overset{''}{33}$
Time in the General Table for ⎱ that Anomaly - - ⎰	2 . 8631776	Days. 729 . 7558
$D^{\frac{3}{2}}$ - - -	8 . 5957789	
Time from the Perihelion	1 . 4589565	28 . 7711

	DAYS.
Epoch September - -	8 . 58333
Time from the Perihelion - -	28 . 7711
Time of the Comet's paffing the Perihelion October	7 . 35443

Application of Mr. De La Place's Method for correcting

the Orbit of a Comet, to the Comet of 1769.

§ 1. THIS Example is taken from the *Cometographie* of Mr. *Pingrè*; with this difference; that the Procefs for the Solution of the Triangles in the Firft Hypothefis, is given at full length. The operation in the other Two Hypothefes being perfectly fimilar; it would have been fuperfluous to give them in equal detail. The Approximate Time of Arrival at the Perihelion, and Perihelion diftance, are nearer the Truth than thofe which have been determined by the Obfervations made before the Perihelion, in the former Part of this Work. As however the Example will be equally inftructive what-ever Elements are ufed; I thought it not worth while to reject the Example furnifhed by fo able a Computer, on that Account.

§ 2. Let the Approximate Perihelion Diftance of the Comet of 1769, be   -   -   0 . 1231459

And the Approximate Time of its Arrival at the Perihelion be   -   -   -   *October* 7^{D} . 54438.

Z z

For the Correction of thefe Elements, let the following Obfervations be chofen.

Times of Obfervation in Days and Decimals.	Longitudes of the Sun.	Obferved Longitudes of the Comet.	Obferved Latitudes of the Comet.	Logarithms of the Sun's diftance from the Earth.	Elongations of the Comet. ɛ, ɛ', and ɛ".
D Auguſt 14 . 52352	° ′ ″ 142 . 21 . 26	° ′ ″ 39 . 58 . 16	° ′ ″ — 3 . 17 . 13	0 . 005244	° ′ ″ ♎ 102 . 23 . 10
Septem. 15 . 69398	173 . 31 . 3c	140 . 39 . 17	—22 . 43 . 34	0 . 001810	♎ 32 . 52 . 13
Decem. 2 . 21413	250 . 54 . 12	276 . 41 . 20	+23 . 33 . 2⁊	9 . 993496	♓ 25 . 47 . 8

The South Latitudes of the Comet are marked with the Negative Sign; the North, with the Affirmative.

Firſt Hypothefis.

Time of Obfervation -	Auguſt	September	December
	D 14 . 52352	D 15 . 69398	D 2 . 21413
Interval from Perihelion	54 . 02086	21 . 85040	55 . 66975
Logarithm of Interval	1 . 7325615	1 . 3394594	1 . 7456193
Perihelion Diftance $\frac{3}{2}$ —	8 . 6356298	8 . 6356298	8 . 6356298
Logarithm of Days	3 . 0969317	2 . 7038296	3 . 1099895
Days - -	1250 . 062	505 . 626	1288 . 22
Anomaly -	° ′ ″ 146 . 13 . 24	° ′ ″ 132 . 23 . 36	° ′ ″ 146 . 35 . 35
Half of the Anomaly	73 . 6 . 42	66 . 11 . 48	73 . 17 . 48¼
Cofine of the ½ Anomaly	9 . 4631571	9 . 6059496	9 . 4585148
	2	2	2
Square of the ½ Cofine	8 . 9263142	9 . 2118992	8 . 9170296
Log. of Perihelion Diftance ⎱ from which fubtract ½ Cos² ⎰	9 . 0904199	9 . 0904199	9 . 0904199
Radius Vector Log.	0 . 1641057	9 . 8785207	0 . 1733903

The two firſt Anomalies preceding the Perihelion, are to be conſidered as Negative.   Therefore we have,

$$U = v' - v = \overset{\circ}{13} \cdot \overset{\prime}{49} \cdot \overset{\prime\prime}{48}$$
$$U' = v'' - v = 292 \cdot 48 \cdot 59$$

§ 4. We now proceed to the Solution of the Triangles which from the Obſerved Longitudes and Latitudes of the Comet, and the Radii Veĉtores, give its Heliocentric Longitudes and Latitudes ; and from them, V, and V' ; or the Angular Motions of the Comet as ſeen from the Sun.

§ 5. It may here again be obſerved, that in ſome of the ſubſequent Operations, the Solution of the Triangles may be ambiguous, giving two Angles.   But the Figure uſed for determining the Approximate Elements, if the Abbè Boſcovich's method has been followed, will be a ſure guidè in this Part of the Work.   If Mr. De la Place's method has been uſed, it will be beſt to lay down the Approximate Orbit on Paper to guide theſe Operations.

§ 6. Firſt Triangle $Cos\ \varepsilon \times Cos\ \theta = Cos\ S\ T\ C.$

			Auguſt 14.	*September* 15.	*December* 2.
Cos ε	-	-	9 · 3314244	9 · 9242284	9 · 9544492
Cos θ	-	-	9 · 9992850	9 · 9649015	9 · 9622189
Cos S T C	-		9 · 3307094	9 · 8891299	9 · 9166681
S T C			$\overset{\circ}{102} \cdot \overset{\prime}{21} \cdot \overset{\prime\prime}{55}$	$\overset{\circ}{39} \cdot \overset{\prime}{13} \cdot \overset{\prime\prime}{22}$	$\overset{\circ}{34} \cdot \overset{\prime}{22} \cdot \overset{\prime\prime}{19}$

§ 7. Second Triangle $\dfrac{Sin\ STC \times R}{r} = Sin\ TCS$.

		Auguſt 14.	September 15.	December 2.
Sin STC	-	9 . 9898066	9 . 8009489	9 ; 7517123
R	-	0 . 0052440	0 . 0018100	9 . 9934960
r	a: c:	9 . 8358943	0 . 1214793	9 . 8266097
Sin TCS	-	9 . 8309449	9 . 9242382	9 . 5718180
TCS		42.39.10¼	122.52.6	21.54.23½

§ 8. Thirdly for the Angle $TSC = \overset{\circ}{180} - TCS - STC$.

			Auguſt 14.	September 15.	December 2.
TCS	-	-	42.39.10½	122.52..6	21.54.23½
STC	-	-	102.21.55	39.13.22	34.22.19
			145. 1. 5½	162. 5.28	56.16.42½
TSC	-		34.58.54½	17.54.32	123.43.17½

§ 9. Fourth Triangle $\dfrac{Sin\ \theta \times Sin\ TSC}{Sin\ STC} = Sin\ \varpi$

		Auguſt 14.	September 15.	December 2.
Sin θ	- -	8 . 7584314	9 . 5869544	9 . 6016910
Sin TSC	-	9 . 7583942	9 . 4878511	9 . 9199905
Sin STC	a: c:	0 . 0101934	0 . 1990511	0 . 2482877
Sin ϖ	- -	8 . 5270190	9 . 2738566	9 . 7699692
ϖ	- -	1 . 55 . 43	10 . 49 . 42½	36 . 4 . 19

§ 10. Fifth Triangle $\dfrac{Cos\ TSC}{Cos\ \varpi} = Cos\ TSP.$

	August 14.	*September* 15.	*December* 2.
Cos T S C	9 . 9134611	9 . 9784301	9 . 7444159
Cos ϖ	9 . 9997539	9 . 9921973	9 . 9075609
Cos T S P	9 . 9137072	9 . 9862328	9 . 8368550
T S P	34 . 56 . 7	14 . 21 . 2	133 . 22 . 49

§ 11. The Heliocentric Angles between the Earth and Comet being now found ; the infpection of the Figure will fhew whether the Comet is to the Eaft, or the Weft of the Earth. In the Firft Cafe the Angle T S P, (commonly called the Angle of Commutation) muft be added to the Longitude of the Earth ; in the latter, it muft be fubtracted from it. In the Example before us the Comet was to the Eaft of the Earth at the Firft and Second Obfervations, and to the Weft of it in the Third. Therefore

	August 14.	*September* 15.	*December* 2.
Longitude of the Earth	322 . 21 . 26	353 . 31 . 30	70 . 54 . 12
T S P	34 . 56 . 7	14 . 21 . 2	133 . 22 . 49
$\mathfrak{E}$	357 . 17 . 33	7 . 52 . 32	297 . 31 . 23

§ 12. Having now the $\mathfrak{E}$, $\mathfrak{E}'$ and $\mathfrak{E}''$; or the Three Heliocentric Longitudes of the Comet ; we proceed to find the

3 A

Angular Motions of the Comet in Longitude, between the First and Second Obfervation, and between the Firft and Third Obfervation ; or $\mathfrak{E}'-\mathfrak{E}$, and $\mathfrak{E}''-\mathfrak{E}$. We therefore fay

$\mathfrak{E}'$	$7 \cdot 52 \cdot 32$	$\mathfrak{E}''$	$297 \cdot 31 \cdot 23$
$\mathfrak{E}$	$357 \cdot 17 \cdot 33$	$\mathfrak{E}$	$357 \cdot 17 \cdot 33$
$\mathfrak{E}'-\mathfrak{E}$	$10 \cdot 34 \cdot 59$	$\mathfrak{E}''-\mathfrak{E}$	$300 \cdot 13 \cdot 50$

§ 13. Having now the Comet's Heliocentric Motions in Longitude between the Firft and Second Obfervations, $\mathfrak{E}'-\mathfrak{E}$; and between the Second and Third Obfervations $\mathfrak{E}''-\mathfrak{E}$; and the Three Heliocentric Latitudes ϖ, ϖ', and ϖ'' ; we proceed to find the Heliocentric Angular Motions of the Comet V and V' by the Formula.

$$Cos\ (\mathfrak{E}'-\mathfrak{E})\ Cos\ \varpi\ Cos\ \varpi' + Sin\ \varpi\ Sin\ \varpi' = Cos\ V.$$

$Cos\ (\mathfrak{E}'-\mathfrak{E})$	$9 \cdot 9925490$		
$Cos\ \varpi$	$9 \cdot 9997539$	$Sin\ \varpi$	$8 \cdot 5270190$
$Cos\ \varpi'$	$9 \cdot 9921973$	$Sin\ \varpi'$	$9 \cdot 2738566$
Log.	$9 \cdot 9845002$	Log.	$7 \cdot 8008756$
Number	$0,\ 96493978$	Number	$0 \cdot 00632231$

$Cos\ (\mathfrak{E}'-\mathfrak{E})\ Cos\ \varpi\ Cos\ \varpi'$	$0 \cdot 96493978$
$Sin\ \varpi\ Sin\ \varpi'$	$0 \cdot 00632231$
$Cos\ V$	$0 \cdot 97126209$
Its Logarithm	$9 \cdot 9873364$
	$\overset{\circ}{} \quad ' \quad ''$
V	$13 \cdot 46 \cdot 9$
U	$13 \cdot 49 \cdot 48$
$U-V=m$	$+ \quad 0 \cdot 3 \cdot 39$

§ 14. The Angular Motion of the Comet between the Firſt and Third Obſervations $= V'$, will be found in like Manner by the Formula.

$$\text{Cos } (\mathfrak{C}'' - \mathfrak{C}) \text{ Cos } \varpi \text{ Cos } \varpi'' + \text{Sin } \varpi \text{ Sin } \varpi'' = V'$$

Cos ($\mathfrak{C}''-\mathfrak{C}$)	9 . 7019829		
Cos ϖ	9 . 9997539	Sin ϖ	8 . 5270190
Cos ϖ''	9 . 9075609	Sin ϖ''	9 . 7699692
Log.	+ 9 . 6092977	Log.	— 8 . 2969882
Number	+ 0 . 40672206	Number	— 0 . 01981472

Cos ($\mathfrak{C}''-\mathfrak{C}$) Cos ϖ Cos ϖ''	+ 0 . 40672206
Sin ϖ Sin ϖ''	— 0 . 01981472
Cos V'	+ 0 . 38690734
Its Logarithm	9 . 5876069
V'	292 . 45 . 44
U'	292 . 48 . 59
$U' - V' = n$	+ 0 . 3 . 15

§ 15. As the U and U' are not equal to the V and V'; that is as the Angular Motion of the Comet found directly by its Anomalies; is not the ſame as that deduced from the Geocentric Longitudes and Latitudes obſerved, its aſſumed diſtances from the Sun being the Radii Vectores belonging to thoſe Anomalies, in the Aſſumed Parabola, and Time of Arrival at the Perihelion in that Parabola; it is evident that this Hypotheſis is defective; another therefore muſt be tried.

Second Hypothesis.

§ 16. Leaving the Time of the Comet's Arrival at the Perihelion unaltered, we change the Perihelion Distance ; and as the Errors m, and n, of the First Hypothesis are but small, the Alteration of the Perihelion Distance should be inconsiderable ; we therefore increase it only by o, oo1.

The Elements now are,

Time of Passage by the Perihelion, October 7 . 54438
Perihelion Distance   o . 1241459

As the Process is precisely the same as in the First Hypothesis it would be superfluous to give it at large. The Result is

V	13 . 52 . 11	V'	292 . 51 . 7
U	13 . 54 . 22	U'	292 . 31 . 2
$U—V=m'+$	o . 2 . 11	$U—V'=n'—$	o . 20 . 5

This Second Hypothesis is therefore defective, and a Third must be formed.

Third Hypothesis.

§ 17. We now assume the Perihelion Distance as in the First Hypothesis, but change the Time of the Arrival at the Perihelion ; and as in the Second Hypothesis, by a small quantity only. Viz. by — o . o1 of a Day. The Elements now are,

Time of Passage by the Perihelion, October 7 . 53438
Perihelion Distance   o . 1231459

With thefe Elements we have,

$$
\begin{array}{ll}
V & 13 \cdot 48 \cdot \overset{''}{12} \\
U & \underline{13 \cdot 50 \cdot 11} \\
U - V = m'' + & 0 \cdot 1 \cdot 59
\end{array}
\qquad
\begin{array}{ll}
V' & 292 \cdot \overset{'}{45} \cdot \overset{''}{49} \\
U' & \underline{292 \cdot 49 \cdot 4} \\
U' - V' = n'' + & 0 \cdot 3 \cdot 15
\end{array}
$$

§ 18. This Third Hypothefis being alfo defective, we proceed to determine the true Perihelion Diftance and Time of Arrival from the Errors of the Three Hypothefes.

Collecting them therefore together we have,

$$
\begin{array}{ll}
m & + \ \overset{''}{219} \\
m' & + \ 131 \\
m'' & + \ 119 \\[4pt]
\hline
m - m' & + \quad \overset{''}{88} \\
m - m'' & + \ 100
\end{array}
\qquad\qquad
\begin{array}{ll}
n & + \ \overset{''}{195} \\
n' & - \ 1205 \\
n'' & + \ 195 \\[4pt]
\hline
n - n' & + \quad \overset{''}{1400} \\
n - n'' & \quad\ 0
\end{array}
$$

§ 19. For the Solution therefore of the Equation.

$$
u = \frac{m(n - n'') - n(m - m'')}{(m - m')(n - n'') - (m - m'')(n - n')}
$$

We have,
$$
\begin{array}{lr}
m & + \ 219 \\
n - n'' & 0 \\[2pt]
\hline
m(m - n'') & 0
\end{array}
\qquad
\begin{array}{lr}
n & + \quad 195 \\
m - m'' & + \quad 100 \\[2pt]
\hline
n(m - m'') & + \ 19500
\end{array}
$$

$$
\begin{array}{lr}
m - m' & + \ 88 \\
n - n'' & 0 \\[2pt]
\hline
(m - m')(n - n'') & 0
\end{array}
\qquad
\begin{array}{lr}
m - m'' & + \quad 100 \\
n - n' & + \quad 1400 \\[2pt]
\hline
(m - m'')(n - n') & + \ 140000
\end{array}
$$

3 B

The Numerator therefore of the Equation is   19500
The Denominator   -   -   -   140000

Logarithm of	19500	4 . 2900346
Logarithm of	140000	5 . 1461280
Logarithm of	u	9 . 1439066
	u	0 . 1392857

It may be here obferved that the Term $n - n''$ being 0, deftroys the Terms multiplied by it, and thereby reduces the Equation in this Example to a very fimple Form.

§ 20. Having found u; we proceed to find t, by the folution of the Equation

$$t = \frac{m - u \ (m - m')}{m - m''}$$

Logarithm of	u	9 . 1439066
Logarithm of	$(m - m')$	1 . 9444827
Logarithm of	$u \ (m - m')$	1 . 0883893
	$u \ (m - m)$	+ 12 . 25715

m	+	219 .
$u \ (m - m')$	+	12 . 25715
	+	206 . 74285

This therefore is the Numerator of the Equation, and its Denominator $m - m''$ being + 100, the divifion is fimply placing the Decimal point Two places forward, and we have,

$$t = \quad + 2 . 0674285.$$

§ 21. It now remains to multiply the change in the Perihelion Diſtance, and Time of Arrival at the Perihelion, by u, and t, reſpectively, in order to obtain the true Variation of each from theſe Elements, as aſſumed in the Firſt Hypotheſis.

Change of Perihelion Diſtance	$+$ 0 . 001
u	$+$ 0 . 1392857
True Change	$+$ 0 . 0001393
Perihelion Diſtance aſſumed	0 . 1231459
True Perihelion Diſtance	0 . 1232852
Change of Time of Arrival	$-$ 0 . 01
t	$+$ 2 . 0674285
True Change	$-$ 0 . 02067
Time of Arrival aſſumed October	7 . 54438
True Time of Arrival at the Perihelion October	7 . 52371

		D	H	M	S
Or	October	7 .	12 .	34 .	9

§ 22. As a Proof of the Accuracy of the preceding Operations, a Fourth Hypotheſis may be formed with the Perihelion Diſtance, and Time of Arrival at the Perihelion, laſt determined ; (and indeed this Operation is neceſſary for finding the Heliocentric Longitudes and Latitudes of the Comet, which are the Elements of the Equations for the Line of Nodes and Inclination ;) in this Hypotheſis we have,

	Auguſt 14.	*September* 15.	*December* 2.
Heliocentric Latitude	$-$ 1 . 55 . 39	$-$10 . 52 . 16	$+$36 . 4 . 9½
Heliocentric Longitude	357 . 16 . 26	7 . 56 . 6	297 . 31 . 1
V	13 . 51 . 22	V'	292 . 46 . 31
U	13 . 51 . 12	U'	292 . 46 . 31
Difference	0 . 0. 10	Difference	0 . 0 . 0

The Difference between V' and U' is nothing. That between V and U is fo fmall, that unlefs the whole Operation had been carried to the precifion of Decimals of Seconds, a lefs could not be expected.

Determination of the Remaining Elements of the Orbit.

§ 24. Having the Heliocentric Longitudes and Latitudes of the Comet, we proceed to find the Place of the Afcending Node by the Solution of the following Equation.

$$\text{Tangent } N = \frac{\text{Tang } \varpi'' \; \text{Sin } \mathfrak{C} - \text{Tang } \varpi \; \text{Sin } \mathfrak{C}''}{\text{Tang } \varpi'' \; \text{Cos } \mathfrak{C} - \text{Tang } \varpi \; \text{Cos } \mathfrak{C}''}$$

N being the Place of the Afcending Node.

Tang ϖ''	$+$	9 . 8623653	Tang ϖ	$-$	8 . 5270357
Sin $\mathfrak{C}$	$-$	8 . 6772570	Sin $\mathfrak{C}''$	$-$	9 . 9478620
	$-$	8 . 5396223		$+$	8 . 4748977
Number	$-$	0 · 03464354	Number	$+$	0 . 02984680
Subtract	$+$	0 · 02984680			
Remainder	$-$	0 . 06449034	$=$ Tang ϖ'' Sin $\mathfrak{C}$ $-$ Tang ϖ Sin $\mathfrak{C}''$		

Tang ϖ''	$+$	9 . 8623653	Tang ϖ	$-$	8 . 5270357
Cos $\mathfrak{C}$	$+$	9 . 9995082	Cos $\mathfrak{C}''$	$+$	9 . 6646522
		9 . 8618735			8 . 1916879
Number	$+$	0 . 72756790	Number	$-$	0 . 0155485
Subtract	$-$	0 . 01554850			
Remainder	$+$	0 . 74311640	$=$ Tang ϖ'' Cos $\mathfrak{C}$ $-$ Tang ϖ Sin $\mathfrak{C}''$		

$$\text{Tang } \varpi'' \text{ Sin } \mathstrut 6 - \text{Tang } \varpi \text{ Sin } 6'' \text{ Log.} \qquad - \quad 8 \cdot 8094947$$
$$\text{Tang } \varpi'' \text{ Cos } 6 - \text{Tang } \varpi \text{ Cos } 6'' \text{ Log.} \quad + \quad 9 \cdot 8710569$$
$$\text{Log. Tangent } N \qquad\qquad\qquad\qquad\qquad - \quad 8 \cdot 9384378$$

This Tangent may belong to Two Angles; but the Approximate Elements before obtained, determine that the Real Angle is $175 \cdot 2 \cdot 24$. This therefore is the Longitude of the Afcending Node of the Comet's Orbit.

§ 25. Having the Longitude of the Afcending Node, the Inclination is given by the Equation.

$$\text{Tang } I = \frac{\text{Tang } \varpi''}{\text{Sin } (6'' - N)}$$

I being the Inclination of the Orbit of the Comet to the Ecliptic. We have,

$6''$	$297 \cdot 31 \cdot 1$
N	$175 \cdot 2 \cdot 24$
$6'' - N$	$122 \cdot 28 \cdot 37$

$\text{Tang } \varpi''$	$9 \cdot 8623653$
$\text{Sin } (6'' - N)$	$9 \cdot 9261405$
$\text{Tang } I$	$9 \cdot 9362248$
I	$40 \cdot 48 \cdot 29$

§ 26. Let us now fuppofe, at any of the Obfervations, (the Third is now the moft proper) a Spherical Right Angled

3 C

Triangle in the Heavens; one of whose sides is the Heliocentric Latitude of the Comet ϖ''; the other, the Arch of the Ecliptic between the Heliocentric Longitude of the Comet at that Observation, and the Ascending Node, $\mathcal{C}'' - N$; then will the Hypothenuse of that Triangle, be the Distance of the Comet from its Node at the Observation, measured on the Orbit of the Comet. This Distance will be found by the Equation.

$$Cos \,\psi = Cos \,(\mathcal{C}'' - N) \; Cos \,\varpi''$$

ψ being the Distance of the Comet from the Node, on the Orbit.

We have therefore

Cos $(\mathcal{C}'' - N)$	9 . 7299421
Cos ϖ''	9 . 9075754
Cos ψ	9 . 6375175
ψ	115 . 43 . 24

§ 27. We have now the Distance of the Comet from the Ascending Node at the Third Observation, ψ; and also its Distance from the Perihelion equal to its Anomaly, at the same Observation, v''; the difference between these, is the Distance between the Place of the Perihelion, and the Ascending Node, measured on the Orbit of the Comet.

v''	146 . 34 . 40
ψ	115 . 43 . 24
Distance from Node to Perihelion	30 51 . 16

Now as the Latitude of the Comet was North, it had paffed the Afcending Node. It had alfo paffed the Perihelion; and as its Diftance from the Perihelion, is greater than from the Afcending Node; it had paffed the Perihelion, before it paffed the Node. And as the Motion of the Comet is direct, or according to the Order of the Signs, the Longitude of the Perihelion muft be lefs than that of the Node; therefore from the Longitude of the Afcending Node, fubtract the Diftance between the Node and Perihelion, and we fhall have the Longitude of the Perihelion.

	$\overset{\circ}{175}$. $\overset{\prime}{2}$. $\overset{\prime\prime}{24}$
N	175 . 2 . 24
Diftance from Node to Perihelion	30 . 51 . 16
Longitude of the Perihelion	144 . 11 . 8
Meafured on the Orbit of the Comet.	

§ 28. Collecting therefore into one view the Refult of the Calculations, we fhall find that from the Obfervations made on the Comet of 1769, on Auguft 14, September 15, and December 2, the Elements of its Orbit will be as follow.

	S . $^{\circ}$. $'$. $''$
Longitude of the Afcending Node	5 . 25 . 2 . 24
Inclination of the Orbit	40 . 48 . 29
Longitude of the Perihelion on the Orbit	4 . 24 . 11 . 8
Perihelion Diftance	0 . . 1232852
Its Logarithm	9 . 0909110

	D . H . M . S
Time of paffage by the Perihelion; October	7 . 12 . 34 . 9
Motion	Direct.

Explanation and Ufe of the Tables.

§ 1. TABLE 1, is for the reduction of Hours, Minutes, and Seconds of Time, into Decimal Parts of a Day.

R U L E.

Find in the Column of Time, the Hours, Minutes, and Seconds given. Oppofite to each is the correfponding Decimal, the Sum of thefe is the Decimal Fraction required.

E X A M P L E.

	H M S
Required the Decimal of	17 . 27 . 44
17 Hours	0 . 708333
27 Minutes	0 . 018750
44 Seconds	0 . 000509
Sum. Decimal of 17 H . 27 M . 44 •	0 . 727592

§ 2. Table 2, is for the Reduction of Decimal Parts of a Day, into Hours, Minutes, and Seconds.

R U L E.

Enter the Table with the Firft Place to the Left Hand, of the given Decimal, and take out its Value in Hours, &c.

3 D

Repeat the fame Operation with the Second, Third, and the Reft of the Places. The Sum of the Times fo taken from the Table, is the Value of the Decimal required.

E X A M P L E.

Required the Value in Time of   0 . 727592

	H	M	s
0 . 7	16 .	48 .	0
0 . 02	0 .	28 .	48
0 . 007	0 .	10 .	4,8
0 . 0005	0 .	0 .	43,2
0 . 00009	0 .	0 .	7,776
0 . 000002	0 .	0 .	0,173
Sum. Value of 0 . 727592	17 .	27 .	43,949

§ 3. Table 3, of the Motion of Comets in a Parabolic Orbit, was firft publifhed by Dr. Halley, and fince augmented by La Caille, De La Lande and Schulze of Berlin. Mr. Pingrè recomputed and extended the whole, fo as to make it much more complete than any before publifhed. And lately Mr. De Lambre, whofe Abilities as a Calculator are well known, has recomputed the whole Table to Decimals of Seconds, and ftill farther enlarged it. While yet unpublifhed he moft liberally communicated it for infertion in this Work; and from his Manufcript it is now printed. Its Ufes are as follow.

§ 4. The Perihelion Diftance of any Comet and the Time of its Paffage by the Perihelion being given, its true Anomaly or Angular Diftance from the Perihelion for any given Time before or after the Perihelion, is required.

R U L E.

To the Logarithm of the Perihelion Diſtance of the Comet, add its half. Subtract the Sum from the Logarithm of the Time elapſed between the given Time and the Arrival of the Comet at its Perihelion; (which by Table 1, muſt be reduced to Days and Decimals) the Remainder will be the Logarithm of a Number of Days and Decimals. Find this Number in the Table of the Parabola, and oppoſite to it is the Anomaly ſought. If the given Number is not in the Table, a ſimple Proportion will give the Anomaly.

If the Characteriſtic of the Logarithm of the Perihelion Diſtance be 9, 8, or 7, in taking its Half, it muſt be ſuppoſed 19, 18, or 17.

E X A M P L E.

The Logarithm of the Perihelion Diſtance of the Comet of 1769 was according to Euler 9 . 0886320. What was its Anomaly at 50 Days before or after its Perihelion?

Log. of Perihelion Diſtance	9 . 0886320
Its Half	9 . 5443160
Their Sum	8 . 6329480
Log. of 50 Days	1 . 6985700
Subtract the above Sum	8 . 6329480
Remainder	3 . 0656220

Which is the Logarithm of 1164 . 185 Days. Seek this Number in the Table of the Parabola. It is not there, but for

1160 Days the Anomaly is 145 . 16 . 49. and for 1165 Days the Anomaly is 145 . 20 . 7. The Difference for 5 Days is 3 . 18. Then fay,

As 5 Days : 3 . 18 : : 4, 185 Days : 2 . 46, which muſt be added to the Anomaly for 1160 Days. 145 . 16 . 49 and the Sum 145 . 19 . 35 is the true Anomaly for 1164, 185 Days in the Table, or 50 Natural ays for the Comet of 1769 before or after its Paſſage by the Perihelion.

§ 5. If the true Anomaly of a Comet for any Moment, be given, and the Time elapſed between that and the Paſſage by the Perihelion is required, the ſame Table may be uſed in an inverſe method.

R   U   L   E.

Seek in the Table the given Anomaly, and find the Time correſponding to it, taking if neceſſary, proportional Parts. To the Logarithm of the Perihelion Diſtance add its Half, and the Logarithm of the Days found in the Table ; their Sum is the Logarithm of the Time elapſed between the Comet's paſſing the Perihelion, and its Arrival at the Anomaly given.

E   X   A   M   P   L   E.

Given the Anomaly of Halley's Comet of 1759, 64 . 36 . 37. Required the Time it took to move that Angle from the

Perihelion. The Logarithm of its Perihelion Diſtance was
9 . 766033 according to the Abbè De La Caille.

The given Anomaly is not in the Table; the Two neareſt
are 64 . 29 . 47, and 64 . 40 . 28. The Firſt anſwers to
58 . 75 Days, and the other to 59 . 0 Days. The Difference
of the given Anomaly from the firſt of theſe Two Tabular ones
is 6 . 50 or 410; the Difference of the Tabular Anomalies
is 10 . 41 or 641, and the Difference of Time is 0 . 25
Days. Then ſay,

$$641 : 0 . 25 :: 410 : 0 . 15991,$$

0 . 15991 muſt therefore be added to the Tabular Time
58 . 75 Days, anſwering to the Anomaly 64 . 29 . 47: the
Sum 58, 90991 Days, will be the Tabular Time anſwering
to the given Anomaly. Now the

Logarithm of 58 . 90991	1 . 7701883
Logarithm of Perihelion Diſtance	9 . 7660330
Half Logarithm of Perihelion Diſtance	9 . 8830165
Sum	1 . 4192378

Whoſe Number is 26 . 25656, the Number of Days that
the Comet will employ in moving the Angle of 64 . 36 . 37
on either ſide of the Perihelion.

§ 6. This general Table will be ſufficient in all Caſes to
determine the True Anomaly from the Time given; but it

3 E

will not be equally accurate for finding the Time from the nomaly for at confiderable Diftances from the Perihelion Earors will arife. The following little Table fhews how far the Table may be ufed without incurring an Error greater than 30 Seconds of Time.

Perihelion Diftance	Anomalies
	D
0 .. 25	130
0 . 50	118
0 . 80	100
1 . 00	90
1 . 20	80
1 . 50	65
2 . 00	50

Beyond thefe Anomalies Comets of the refpective Perihelion Diftances are feldom vifible, and for Comets of a lefs Perihelion Diftance, the Limits extend proportionably further. Indeed, except when extreme accuracy is required, this Table may be ufed far beyond the Limits here prefcribed; and if the utmoft precifion is neceffary, the following Rule will give the Time free from Error, in all Cafes. The Demonftration. will be found in Pingrè, Vol. 2. Page 339.

To the Log. Tangent of Half the given Anomaly, add the conftant Logarithm 1, 9149328; and to triple the Log. Tangent of the Half Anomaly, add the conftant Logarithm 1, 4378116; find the Numbers of thefe Logarithms and add them together. To the Logarithm of the Sum, add ½ of the Logarithm of the Perihelion Diftance of the Comet. The Sum will be the Logarithm of the Days from the Perihelion.

E X A M P L E.

Required the Time from the Perihelion, anſwering to

$\overset{\circ}{144}$. 38 . $\overset{''}{28}$ Anomaly for the Comet of 1769; its Perihelion Diſtance being 9, 0886320.

	LOGARITHMS.	NUMBERS.
Tang. ½ Anomaly -	0 . 4965560	
Conſtant - -	1 . 9149328	
Sum - -	2 . 4114888	257, 92225
Triple Tang. ½ Anomaly	1 . 4896680	
Conſtant - -	1 . 4378116	
Sum - -	2 . 9274796	846, 21275
Sum for the Comet of 109 Days	3 . 0430222	1104, 13500
Perihelion Diſtance ³⁄₂	8 . 6329480	
Time in Days from the Perihelion	1 . 6759702	

§ 7. Comets of extremely ſmall Perihelion Diſtance, ſuch as the Comet of 1680, will be viſible at Diſtances from the Perihelion greater than the Limits of this Table. No Comet has yet been ſeen to which this Obſervation applies, except that of 1680: however in ſuch a caſe, the next Table will ſupply the Deficiency of this; as it extends to the utmoſt poſſible Limits, and with a very ſmall Error in the moſt extreme Caſes.

§ 8. Table 4, is likewiſe for finding the Anomaly, and Radius Vector of a Comet; from the Perihelion Diſtance, and Time of Arrival given, and the contrary. This excellent Table was computed by Mr. Barker, and publiſhed by him in a very ingenious Tract, entitled " An Account of the Diſco-

" veries concerning Comets," printed in the Year 1757. The Rules for ufing it are as follow.

§ 9. Given the Perihelion Diftance of a Comet; and the Time of its Arrival at the Perihelion : Required its True Anomaly, and Radius Vector, for any inftant before or after the Time of Arrival at the Perihelion.

RULE FOR THE ANOMALY.

To the Logarithm of the Perihelion Diftance add its Half; Subtract the Sum, from the conftant Logarithm 9 . 9601283, (which is the Logarithm of the Diurnal Motion of the Comet whofe Perihelion Diftance is 1 . 00) the Remainder is the Logarithm of the Diurnal Motion of the Comet. To this Logarithm add the Logarithm of the Days and Decimals of a Day by which the given Time precedes or follows the Time of the Comet's Arrival at the Perihelion; the Sum will be the Logarithm of the Mean Motion. If this Logarithm exceeds 1 . 517428, enter the Table directly; and oppofite the Logarithm of the mean Motion is the Anomaly fought. If the exact Logarithm is not in the Table; fay, as the Difference in the Table, is to 5; fo is the difference of the given Logarithm from the neareft to it, but lefs than it; to the Minutes and Seconds of Anomaly to be added to the Anomaly oppofite that Logarithm to give the true Anomaly.

If the Logarithm is lefs than 1 . 517428, then its Natural Number muft be found, with which the Table muft be entered under the Title Mean Motion; and a proportion made, if the exact Number be not in the Table as before directed.

RULE FOR THE RADIUS VECTOR.

With the Anomaly enter the Column entitled Logarithms of Diftance, and if the exact Anomaly is not in the Table, fay,

as 5, to the Tabular Difference; fo are the Minutes and Seconds by which the given Anomaly exceeds the Tabular Anomaly nearest to it; to a Number, which added to the Logarithm of Diftance of that Tabular Anomaly, gives the True Logarithm of Diftance. To this Logarithm, add the Logarithm of the Perihelion Diftance, the Sum is the Logarithm of the Radius Vector.

E X A M P L E.

Given the Perihelion Diftance of the Comet of 1769 = 𝜏 0 . 12272, its Logarithm 9 . 0889150. Its Paffage by the Perihelion October 7 . 13 . 58 . 40 $\overset{\text{D H M S}}{}$ or in Decimals 7 . 58241 $\overset{\text{D}}{}$ Required its Anomaly and Radius Vector Auguft 21 . 13 . 4 . 53 $\overset{\text{D H M S}}{}$ or in Decimals 21 . 54506.

As the given Time precedes the Time of the Comet's Paffage by the Perihelion; it muft be fubtracted from that Time, therefore,

	D
From October	7 . 58241
Subtract Auguft	21 . 54506
The Remainder	47 . 03735

is the Comet's Diftance from the Perihelion, in Days and Decimals of a Day.

3 F

Then for the Logarithm of the Diurnal Motion of the Comet.

Log. of Perihelion Diſtance	9 . 0889150
Its Half	9 . 5444575
π ¾ Log.	8 . 6333725
Log. of Diurnal Motion of the Comet of 109 Days conſtant }	9 . 9601283
π ¾ Log. Subtract	8 . 6333725
Log. of Diurnal Motion of Comet	1 . 3267558
Log. of Days from Perihelion add	1 . 6724428
Log. of Mean Motion	2 . 9991986
Neareſt Tabular Log.	2 . 9984800
Difference	719

Then 2756 : 3̈00 :: 719 : 7̈8, 2, to be added.

Anomaly of Tabular Log.	144 . 3̇0 . 0̈0
Proportional Anomaly	0 . 1 . 18, 2
True Anomaly of the Comet	144 . 31 . 18, 2

For the RADIUS VECTOR.

Log. of Diſtance for 144 . 3̇0 . 0̈	1 . 031787
Then 3̈00 : 1976 :: 7̈8, 2 : 515 to be added	515
Log. of Diſtance for 144 . 31 . 1̈8, 2	1 . 032302
Log. of Perihelion Diſtance	9 . 088915
Log. of Radius Vector	0 . 121217

By Table 1, the Anomaly is precisely the same; and the Log. of the Radius Vector is 0 121215. an insensible Difference.

§ 10. If from the Perihelion Distance and Anomaly of a Comet given; it be required to find the Distance of the Comet from the Perihelion in Time; the following Rule will give it.

I. Find the Logarithm of the Diurnal Motion of the Comet as directed in the last §.

II. With the given Anomaly, enter the Column of Mean Motion, and by a proportion if necessary, find the Tabular Mean Motion, or its Logarithm, answering to the given Anomaly.

III. From the Log. of the Mean Motion, subtract the Log. of Diurnal Motion; the Remainder is the Log. of the Comet's Distance from the Perihelion in Days and Decimals.

§ 11. From the Perihelion Distance and Radius Vector, given; the Anomaly may be thus found. From the Logarithm of the Radius Vector, subtract the Logarithm of the Perihelion Distance. With the Remainder enter the Column of Log. of Distance; and (proportioning if necessary) opposite to it is the Anomaly, whence the Time may be found, as before.

§ 12. The Anomaly and Radius Vector being given, the Perihelion Distance is found by the following short Process.

From the Logarithm of the Radius Vector, ſubtract the Log. of Diſtance anſwering to the given Anomaly. The Remainder is the Log. of the Perihelion Diſtance.

Theſe Three laſt Rules ſeem ſo eaſy as to require no Examples.

TABLES

FOR

CONVERTING TIME INTO DECIMALS OF A DAY;

OF THE

PARABOLA;

OF THE

PARABOLIC FALL OF COMETS;

AND

BARKER's TABLE OF THE MEAN MOTION AND RADII
VECTORES OF

COMETS:

WITH

RULES FOR THE USE OF THE TABLES.

TABLE I. *For converting Time into Decimals of a Day.*

Hours.	DECIMALS.	Minutes.	DECIMALS.	Minutes.	DECIMALS.
1	0,04166 . . .	1	0,000694 . . .	31	0,021527 . . .
2	0,08333 . . .	2	0,001388 . . .	32	0,022222 . . .
3	0,12500 . . .	3	0,002083 . . .	33	0,022916 . . .
4	0,16666 . . .	4	0,002777 . . .	34	0,023611 . . .
5	0,20833 . . .	5	0,003472 . . .	35	0,024305 . . .
6	0,25000 . . .	6	0,004166 . . .	36	0,025000 . . .
7	0,29166 . . .	7	0,004861 . . .	37	0,025694 . . .
8	0,33333 . . .	8	0,005555 . . .	38	0,026388 . . .
9	0,37500 . . .	9	0,006250 . . .	39	0,027083 . . .
10	0,41666 . . .	10	0,006944 . . .	40	0,027777 . . .
11	0,45833 . . .	11	0,007638 . . .	41	0,028472 . . .
12	0,50000 . . .	12	0,008333 . . .	42	0,029166 . . .
13	0,54166 . . .	13	0,009027 . . .	43	0,029861 . . .
14	0,58333 . . .	14	0,009722 . . .	44	0,030555 . . .
15	0,62500 . . .	15	0,010416 . . .	45	0,031250 . . .
16	0,66666 . . .	16	0,011111 . . .	46	0,031944 . . .
17	0,70833 . . .	17	0,011805 . . .	47	0,032638 . . .
18	0,75000 . . .	18	0,012590 . . .	48	0,033333 . . .
19	0,79166 . . .	19	0,013194 . . .	49	0,034027 . . .
20	0,83333 . . .	20	0,013888 . . .	50	0,034722 . . .
21	0,87500 . . .	21	0,014583 . . .	51	0,035416 . . .
22	0,91666 . . .	22	0,015277 . . .	52	0,036111 . . .
23	0,95833 . . .	23	0,015972 . . .	53	0,036805 . . .
24	1,00000 . . .	24	0,016666 . . .	54	0,037500 . . .
		25	0,017361 . . .	55	0,038194 . . .
		26	0,018055 . . .	56	0,038888 . . .
		27	0,018750 . . .	57	0,039583 . . .
		28	0,019444 . . .	58	0,040277 . . .
		29	0,020138 . . .	59	0,040972 . . .
		30	0,020833 . . .	60	0,041666 . . .

TABLE I. *Continued.*

Seconds.	DECIMALS.	Seconds.	DECIMALS.
1	0,00001157	31	0,00035880
2	0,00002315	32	0,00037037
3	0,00003472 . . .	33	0,00038194 . . .
4	0,00004630	34	0,00039352
5	0,00005787	35	0,00040509
6	0,00006944 . . .	36	0,00041666 . . .
7	0,00008102	37	0,00042824
8	0,00009259	38	0,00043981
9	0,00010416 . . .	39	0,00045138 . . .
10	0,00011574	40	0,00046296
11	0,00012731	41	0,00047454
12	0,00013888 . . .	42	0,00048611 . . .
13	0,00015046	43	0,00049769
14	0,00016204	44	0,00050926
15	0,00017361 . . .	45	0,00052083 . . .
16	0,00018518	46	0,00053241
17	0,00019676	47	0,00054398
18	0,00020833 . . .	48	0,00055555 . . .
19	0,00021991	49	0,00056713
20	0,00023148	50	0,00057870
21	0,00024305 . . .	51	0,00059027 . . .
22	0,00025463	52	0,00060185
23	0,00026620	53	0,00061343
24	0,00027777 . . .	54	0,00062500 . . .
25	0,00028935	55	0,00063657
26	0,00030093	56	0,00064815
27	0,00031250 . . .	57	0,00065972 . . .
28	0,00032407	58	0,00067130
29	0,00033565	59	0,00068287
30	0,00034722 . . .	60	0,00069444 . . .

Hours.	DECIMALS.	Minutes.	DECIMALS.	Minutes.	DECIMALS.
1	0,04166 . . .	1	0,000694 . . .	31	0,021527 . . .
2	0,08333 . . .	2	0,001388 . . .	32	0,022222 . . .
3	0,12500 . . .	3	0,002083 . . .	33	0,022916 . . .
4	0,16666 . . .	4	0,002777 . . .	34	0,023611 . . .
5	0,20833 . . .	5	0,003472 . . .	35	0,024305 . . .
6	0,25000 . . .	6	0,004166 . . .	36	0,025000 . . .
7	0,29166 . . .	7	0,004861 , . .	37	0,025694 . . .
8	0,33333 . . .	8	0,005555 . . .	38	0,026388 . . .
9	0,37500 . . .	9	0,006250 . . .	39	0,027083 . . .
10	0,41666 . . .	10	0,006944 . . .	40	0,027777 . . .
11	0,45833 . . .	11	0,007638 . . .	41	0,028472 . . .
12	0,50000 . . .	12	0,008333 . . .	42	0,029166 . . .
13	0,54166 . . .	13	0,009027 . . .	43	0,029861 . . .
14	0,58333 . . .	14	0,009722 . . .	44	0,030555 . . .
15	0,62500 . . .	15	0,010416 . . .	45	0,031250 . . .
16	0,66666 . . .	16	0,011111 . . .	46	0,031944 . . .
17	0,70833 . . .	17	0,011805 . . .	47	0,032638 . . .
18	0,75000 . . .	18	0,012500 . . .	48	0,033333 . . .
19	0,89166 . . .	19	0,013194 . . .	49	0,034027 . . .
20	0,83333 . . .	20	0,013888 . . .	50	0,034722 . . .
21	0,87500 . . .	21	0,014583 . . .	51	0,035416 . . .
22	0,91666 . . .	22	0,015277 . . .	52	0,036111 . . .
23	0,95833 . . .	23	0,015972 . . .	53	0,036805 . . .
24	1,00000 . . .	24	0,016666 . . .	54	0,037500 . . .
		25	0,017361 . . .	55	0,038194 . . .
		26	0,018055 . . .	56	0,038888 . . .
		27	0,018750 . . .	57	0,039583 . . .
		28	0,019444 . . .	58	0,040277 . . .
		29	0,020138 . . .	59	0,040972 . . .
		30	0,020833 . . .	60	0,041666 . . .

TABLE I. *For converting Time into Decimals of a Day.*

Seconds.	DECIMALS.	Seconds.	DECIMALS.	Seconds.	DECIMALS.
1	0,00001157	21	0,00024305 ...	41	0,00047454
2	0,00002315	22	0,00025463	42	0,00048611 ...
3	0,00003472 ...	23	0,00026620	43	0,00049769
4	0,00004630	24	0,00027777 ...	44	0,00050926
5	0,00005787	25	0,00028935	45	0,00052083 ...
6	0,00006944 ...	26	0,00030093	46	0,00053241
7	0,00008102	27	0,00031250 ...	47	0,00054398
8	0,00009259	28	0,00032407	48	0,00055555 ...
9	0,00010416 ...	29	0,00033565	49	0,00056713
10	0,00011574	30	0,00034722 ...	50	0,00057870
11	0,00012731	31	0,00035880	51	0,00059027 ...
12	0,00013888 ...	32	0,00037037	52	0,00060185
13	0,00015046	33	0,00038194 ...	53	0,00061343
14	0,00016204	34	0,00039352	54	0,00062500 ...
15	0,00017361 ...	35	0,00040509	55	0,00063657
16	0,00018518	36	0,00041666 ...	56	0,00064815
17	0,00019676	37	0,00042824	57	0,00065972 ...
18	0,00020833 ...	38	0,00043981	58	0,00067130
19	0,00021991	39	0,00045138 ...	59	0,00068287
20	0,00023148 .	40	0,00046296	60	0,00069444 ...

TABLE II. *For converting Decimals of a Day into Time.*

Dec.	H. M. S.	Dec.	H. M. S.	Dec.	M. S.	Dec.	M. S.	Dec.	S.
,1	2. 24. 0	,01	0. 14. 24	,001	1 26,4	,0001	0 8,64	,00001	0,864
,2	4. 48. 0	,02	0. 28. 48	,002	2 52,8	,0002	0 17,28	,00002	1,728
,3	7. 12. 0	,03	0. 43. 12	,003	4 19,2	,0003	0 25,92	,00003	2,592
,4	9. 36. 0	,04	0. 57. 36	,004	5 45,6	,0004	0 34,56	,00004	3,456
,5	12. 0. 0	,05	1. 12. 0	,005	7 12,0	,0005	0 43,20	,00005	4,320
,6	14. 24. 0	,06	1. 26. 24	,006	8 38,4	,0006	0 51,84	,00006	5,184
,7	16. 48. 0	,07	1. 40. 48	,007	10 4,8	,0007	1 0,48	,00007	6,048
,8	19. 12. 0	,08	1. 55. 12	,008	11 31,2	,0008	1 9,12	,00008	6,912
,9	21. 36. 0	,09	2. 9. 36	,009	12 57,6	,0009	1 17,76	,00009	7,776

III. General Table of the Parabola, by M. DE LAMBRE.

Days	Anomaly D. M. S.	Differ. M. S.	Days	Anomaly D. M. S.	Differ. M. S.	Days	Anomaly D. M. S.	Differ. M. S.
0,00	0 0 0,0	20 54,5	10,00	13 48 13,4	20 17,7	20,00	26 51 17,3	18 41,4
25	0 20 54,5	54,4	25	14 8 31,1	15,8	25	27 9 58,7	38,4
50	0 41 48,9	54,3	50	14 28 46,9	14,0	50	27 28 37,1	35,5
0,75	1 2 43,2	20 54,2	10,75	14 49 0,9	20 12,2	20,75	27 47 12,6	18 32,5
1,00	1 23 37,4	54,0	11,00	15 9 13,1	10,3	21,00	28 5 45,1	29,5
25	1 44 31,4	53,8	25	15 29 23,4	8,4	25	28 24 14,6	26,5
50	2 5 25,2	53,5	50	15 49 31,8	6,4	50	28 42 41,1	23,5
1,75	2 26 18,7	20 53,2	11,75	16 9 38,2	20 4,3	21,75	29 1 4,6	18 20,4
2,00	2 47 11,9	52,8	12,00	16 29 42,5	2,4	22,00	29 19 25,0	17,3
25	3 8 4,7	52,4	25	16 49 44,9	20 0,3	25	29 37 42,3	14,3
50	3 28 57,1	51,9	50	17 9 45,2	19 58,2	50	29 55 56,6	11,2
2,75	3 49 49,0	20 51,4	12,75	17 29 43,4	56,0	22,75	30 14 7,8	18 8,0
3,00	4 10 40,4	50,8	13,00	17 49 39,4	53,8	23,00	30 32 15,8	5,0
25	4 31 31,2	50,3	25	18 9 33,2	51,7	25	30 50 20,8	18 1,8
50	4 52 21,5	49,6	50	18 29 24,9	49,4	50	31 8 22,6	17 58,6
3,75	5 13 11,1	20 49,0	13,75	18 49 14,3	19 47,2	23,75	31 26 21,2	55,5
4,00	5 34 0,1	48,2	14,00	19 9 1,5	44,9	24,00	31 44 16,7	52,3
25	5 54 48,3	47,4	25	19 28 46,4	42,5	25	32 2 9,0	49,1
50	6 15 35,7	46,6	50	19 48 28,9	40,1	50	32 19 58,1	45,9
4,75	6 36 22,3	20 45,7	14,75	20 8 9,0	19 37,8	24,75	32 37 44,0	17 42,7
5,00	6 57 8,0	44,8	15,00	20 27 40,8	35,4	25,00	32 55 26,7	39,5
25	7 17 52,8	43,9	25	20 47 22,2	32,9	25	33 13 6,2	36,2
50	7 38 36,7	42,8	50	21 6 55,1	30,4	50	33 30 42,4	33,0
5,75	7 59 19,5	20 41,8	15,75	21 26 25,5	19 27,9	25,75	33 48 15,4	17 29,8
6,00	8 20 1,3	40,7	16,00	21 45 53,4	25,4	26,00	34 5 45,2	26,4
25	8 40 42,0	39,6	25	22 5 18,8	22,8	25	34 23 11,6	23,2
50	9 1 21,6	38,4	50	22 24 41,6	20,2	50	34 40 34,8	19,9
6,75	9 22 0,0	20 37,2	16,75	22 44 1,8	19 17,6	26,75	34 57 54,7	17 16,7
7,00	9 42 37,2	35,9	17,00	23 3 19,4	14,9	27,00	35 15 11,4	13,3
25	10 3 13,1	34,7	25	23 22 34,3	12,3	25	35 32 24,7	10,0
50	10 23 47,8	33,3	50	23 41 46,6	9,6	50	35 49 34,7	6,7
7,75	10 44 21,1	20 31,9	17,75	24 0 56,2	19 6,9	27,75	36 6 41,4	17 3,4
8,00	11 4 53,0	30,4	18,00	24 20 3,1	4,1	28,00	36 23 44,8	17 0,0
25	11 25 23,4	29,0	25	24 39 7,2	19 1,3	25	36 40 44,8	16 56,7
50	11 45 52,4	27,5	50	24 58 8,5	18 58,6	50	36 57 41,5	53,4
8,75	12 6 19,9	20 26,0	18,75	25 17 7,1	55,8	28,75	37 14 34,9	50,2
9,00	12 26 45,9	24,4	19,00	25 36 2,9	52,9	29,00	37 31 24,9	46,7
25	12 47 10,3	22,7	25	25 54 55,8	50,0	25	37 48 11,6	43,3
50	13 7 33,0	21,1	50	26 13 45,8	47,2	50	38 4 54,9	40,0
9,75	13 27 54,1	20 19,3	19,75	26 32 33,0	18 44,3	29,75	38 21 34,9	16 36,6

TABLE III. *continued.*

Days.	D.	M.	S.	M.	S.	Days.	D.	M.	S.	M.	S.	Days.	D.	M.	S.	M.	S.
30,00	38	38	11,5	16	33,2	40,00	48	56	34,5	14	19,3	50,00	57	48	20,1	12	15,4
25	38	54	44,7		29,9	25	49	10	53,8		16,0	25	58	0	35,5		12,5
50	39	11	14,6		26,5	50	49	25	9,8		12,7	50	58	12	48,0		9,6
30,75	39	27	41,1			40,75	49	39	22,5			50,75	58	24	57,6		
				16	23,1					14	9,5					12	6,7
31,00	39	44	4,2	16	19,7	41,00	49	53	32,0		6,3	51,00	58	37	4,3		3,9
25	40	0	23,9		16,4	25	50	7	38,3	14	3,0	25	58	49	8,2	12	1,0
50	40	16	40,3		13,0	50	50	21	41,3	13	59,8	50	59	1	9,2	11	58,2
31,75	48	32	53,3			41,75	50	35	41,1			51,75	59	13	7,4		
				16	9,0					13	56,5						55,3
32,00	40	49	2,9		6,3	42,00	50	49	37,6	13	53,4	52,00	59	25	2,7		52,5
25	41	5	9,2	16	2,8	25	51	3	31,0		50,1	25	59	36	55,2		49,7
50	41	21	12,0	15	59,5	50	51	17	21,1		47,0	50	59	48	44,9		46,9
32,75	41	37	11,5			42,75	51	31	8,1			52,75	60	0	31,8		
					56,1					13	43,7					11	44,1
33,00	41	53	7,6		52,8	43,00	51	44	51,8		40,6	53,00	60	12	15,9	11	41,4
25	42	9	0,4		49,3	25	51	58	32,4		37,4	25	60	23	57,3		38,6
50	42	24	49,7		45,9	50	52	12	9,8		34,3	50	60	35	35,9		35,8
33,75	42	40	35,6			43,75	52	25	44,1			53,75	60	47	11,7		
				15	42,6					13	31,1					11	33,1
34,00	42	56	18,2		39,2	44,00	52	39	15,2		27,9	54,00	60	58	44,8		30,4
25	43	11	57,4		35,9	25	52	52	43,1		24,9	25	61	10	15,2		27,6
50	43	27	33,3		32,4	50	53	6	8,0		21,7	50	61	21	42,8		24,9
34,75	43	43	5,7			44,75	53	19	29,7			54,75	61	33	7,7		
				15	29,1					13	18,5					11	22,2
35,00	43	58	34,8		25,7	45,00	53	32	48,2		15,5	55,00	61	44	29,9		19,6
25	44	14	0,5		22,4	25	53	46	3,7		12,4	25	61	55	49,5		16,8
50	44	29	22,9		19,0	50	53	59	16,1		9,3	50	62	7	6,3		14,2
35,75	44	44	41,9			45,75	54	12	25,4			55,75	62	18	20,5		
				15	15,6					13	6,2					11	11,6
36,00	44	59	57,5		12,3	46,00	54	25	31,6		3,1	56,00	62	29	32,1		8,9
25	45	15	9,8		9,0	25	54	38	34,7	13	0,1	25	62	40	41,0		6,3
50	45	30	18,8		5,6	50	54	51	34,8	12	57,0	50	62	51	47,3		3,6
36,75	45	45	24,4			46,75	55	4	31,8			56,75	63	2	50,9		
				15	2,2						54,0					11	1,1
37,00	46	0	26,6	14	58,9	47,00	55	17	25,8		51,0	57,00	63	13	52,0	10	58,4
25	46	15	25,5		55,6	25	55	30	16,8		47,9	25	63	24	50,4		55,9
50	46	30	21,1		52,2	50	55	43	4,7		44,9	50	63	35	46,3		53,3
37,75	46	45	13,3			47,75	55	55	49,6			57,75	63	46	39,6		
				14	48,9					12	41,9					10	50,7
38,00	47	0	2,2		45,6	48,00	56	8	31,5		39,0	58,00	63	57	30,3		48,1
25	47	14	47,8		42,3	25	56	21	10,5		35,9	25	64	8	18,4		45,7
50	47	29	30,1		39,0	50	56	33	46,4		33,0	50	64	19	4,1		43,1
38,75	47	44	9,1			48,75	56	46	19,4			58,75	64	29	47,2		
				14	35,7					12,30,0						10	40,5
39,00	47	53	44,8		32,3	49,00	56	58	49,4		27,1	59,00	64	40	27,7		38,1
25	48	13	17,1		29,1	25	57	11	7,5		24,1	25	64	51	5,8		35,5
50	48	27	46,2		25,8	50	57	23	40,6		21,2	50	65	1	41,3		33,1
39,75	48	42	12,0			49,75	57	36	1,8			59,75	65	12	14,4		
				14	22,5					12	18,3					10	30,6

TABLE III. *continued.*

Days.	D. M. S.	M. S.	Days.	D. M. S.	M. S.	Days.	D. M. S.	M. S.
60,00	65 22 45,0	10 28,1	70,00	71 51 23,2	8 58,3	80,00	77 25 23,1	7 44,3
25	65 33 13,1	25,7	25	72 0 21,5	56,3	25	77 33 7,4	42,5
50	65 43 38,8	23,2	50	72 9 17,8	54,3	50	77 40 49,9	40,9
60,75	65 54 2,0	10 20,8	70,75	72 18 12,1	8 52,3	80,75	77 48 30,8	7 39,3
61,00	66 4 22,8	18,4	71,00	72 27 4,4	50,2	81,00	77 56 10,1	37,6
25	66 14 41,2	16,0	25	72 35 54,6	48,2	21	78 3 47,7	35,9
50	66 24 57,2	13,6	50	72 44 42,8	46,3	50	78 11 23,6	34,3
61,75	66 35 10,8	10 11,2	71,75	72 53 29,1	8 44,3	81,75	78 18 57,9	7 32,7
62,00	66 45 22,0	8,8	72,00	73 2 13,4	42,3	82,00	78 26 30,6	31,1
25	66 55 30,8	6,4	25	73 10 55,7	40,4	25	78 34 1,7	29,5
50	67 5 37,2	4,1	50	73 19 36,1	38,4	50	78 41 31,2	27,8
62,75	67 15 41,3	10 1,8	72,75	73 28 14,5	8 36,5	82,75	78 48 59,0	7 26,3
63,00	67 25 43,1	9 59,4	73,00	73 36 51,0	34,6	83,00	78 56 25,3	24,7
25	67 35 42,5	57,1	25	73 45 25,6	32,6	25	79 3 50,0	23,1
50	67 45 39,6	54,8	50	73 53 58,2	30,8	50	79 11 13,1	21,6
63,75	67 55 34,4	9 52,5	73,75	74 2 29,0	8 28,8	83,75	79 18 34,7	7 19,9
64,00	68 5 26,9	9 50,2	74,00	74 10 57,8	26,9	84,00	79 25 54,6	18,5
25	68 15 17,1	47,9	25	74 19 24,7	25,1	25	79 33 13,1	16,9
50	68 25 5,0	45,6	50	74 27 49,8	23,2	50	79 40 30,0	15,3
64,75	68 34 50,6	9 43,4	74,75	74 36 13,0	8 21,3	84,75	79 47 45,3	7 13,8
65,00	68 44 34,0	41,1	75,00	74 44 34,3	8 19,5	85,00	79 54 59,1	12,3
25	68 54 15,1	38,9	25	74 52 53,8	17,6	25	80 2 11,4	10,7
50	69 3 54,0	36,7	50	75 1 11,4	15,8	50	80 9 22,1	9,3
65,75	69 13 30,7	9 34,4	75,75	75 9 27,2	8 14,0	85,75	80 16 31,4	7 7,7
66,00	69 23 5,1	32,3	76,00	75 17 41,2	12,1	86,00	80 23 39,1	7 6,4
25	69 32 37,4	30,0	25	75 25 53,3	10,3	25	80 30 45,5	4,8
50	69 42 7,4	27,9	50	75 34 3,6	8,5	50	80 37 50,3	3,3
66,75	69 51 35,3	9 25,7	76,75	75 42 12,1	8 6,8	86,75	80 44 53,6	7 1,8
67,00	70 1 1,0	23,5	77,00	75 50 18,9	5,0	87,00	80 51 55,4	7 0,4
25	70 10 24,5	21,3	25	75 58 23,9	3,1	25	80 58 55,8	6 58,9
50	70 19 45,8	19,2	50	76 6 27,0	8 1,4	50	81 5 54,7	57,4
67,75	70 29 5,0	9 17,1	77,75	76 14 28,4	7 59,7	87,75	81 12 52,1	56,0
68,00	70 38 22,1	15,0	78,00	76 22 28,1	57,9	88,00	81 19 48,1	54,6
25	70 47 37,1	12,9	25	76 30 26,0	56,2	25	81 26 42,7	53,2
50	70 56 50,0	10,7	50	76 38 22,2	54,4	50	81 33 35,9	51,7
68,75	71 6 0,7	9 8,6	78,75	76 46 16,6	7 52,7	88,75	81 40 27,6	6 50 3
69,00	71 15 9,3	6,6	79,00	76 54 9,3	51,0	89,00	81 47 17,9	48,9
25	71 24 15,9	4,5	25	77 2 0,3	49,3	25	81 54 6 8	47,5
50	71 33 20,4	2,4	50	77 9 49,6	47,6	50	82 0 54,3	46,1
69,75	71 42 22,8	9 0,4	79,75	77 17 37,2	7 45,9	89,75	82 7 40,4	6 44,8

TABLE III. *continued.*

Days.	D. M. S.	M. S.	Days.	D. M. S.	M. S.	Days.	D. M. S.	M. S.
90,00	82 14 25,2		100,00	86 26 28,7		110,00	90 8 1,3	
25	82 21 8,5	6 43,3	25	86 32 21,9	5 53,2	25	90 13 13,0	5 11,7
50	82 27 50,5	42,0	50	86 38 13,9	52,0	50	90 18 23,8	10,8
90,75	82 34 31,0	40,5	100,75	86 44 4,9	51,0	110,75	90 23 33,6	9,8
		6 39,3			5 49,8			5 8,8
91,00	82 41 10,3	37,8	101,00	86 49 54,7	48,7	111,00	90 28 42,4	8,0
25	82 47 48,1	36,6	25	86 55 43,4	47,5	25	90 33 50,4	7,0
50	82 54 24,7	35,2	50	87 1 30,9	46,5	50	90 38 57,4	6,1
91,75	83 0 59,9		101,75	87 7 17,4		111,75	90 44 3,5	
		6 33,8			5 45,4			5 5,2
92,00	83 7 33,7	32,6	102,00	87 13 2,8	44,2	112,00	90 49 8,7	4,2
25	83 14 6,3	31,2	25	87 18 47,0	43,2	25	90 54 12,9	3,4
50	83 20 37,5	29,9	50	87 24 30,2	42,1	50	90 59 16,3	2,4
92,75	83 27 7,4		102,75	87 30 12,3		112,75	91 4 18,7	
		6 28,5			5 41,0			5 1,6
93,00	83 33 35,9	27,3	103.00	87 35 53,3	39,9	113,00	91 9 20,3	5 0,6
25	83 40 3,2	25,9	25	87 41 33,2	38,8	25	91 14 20,9	4 59,8
50	83 46 29,1	24,7	50	87 47 12,0	37,8	50	91 19 20,7	58,9
93,75	83 52 53,8		103,75	87 52 49,8		113,75	91 24 19,6	
		6 23,4			5 36,7			4 58,0
94,00	83 59 17,2	22,1	104,00	87 58 26,5	35,7	114,00	91 29 17,6	57,1
25	84 5 39,3	20,8	25	88 4 2,2	34,6	25	91 34 14,7	56,2
50	84 12 0,1	19,6	50	88 9 36,8	33,6	50	91 39 10,9	55,3
94,75	84 18 19,7		104,75	88 15 10,4		114,75	91 44 6,2	
		6 18,3			5 32,5			4 54,5
95,00	84 24 38,0	17,1	105,00	88 20 42,9	31,5	115,00	91 49 0,7	53,6
25	84 30 55,1	15,8	25	88 26 14,4	30,4	25	91 53 54,3	52,7
50	84 37 10,9	14,6	50	88 31 44,8	29,4	50	91 58 47,0	52,0
95,75	84 43 25,5		105,75	88 37 14,2		115,75	92 3 39,0	
		6 13,3			5 28,4			4 51,0
96,00	84 49 38,8	12,1	106,00	88 42 42,6	27,4	116,00	92 8 30,0	50,1
25	84 55 50,9	10,9	25	88 48 10,0	26,3	25	92 13 20,1	49,4
50	85 2 1,8	9,6	50	88 53 36,3	25,4	50	92 18 9,5	48,5
96,75	85 8 11,4		106,75	88 59 1,7		116,75	92 22 58,0	
		6 8,4			5 24,3			4 47,7
97,00	85 14 19,8	6 7,3	107,00	89 4 26,0	23,3	117,00	92 27 45,7	46,8
25	85 20 27,1	6,1	25	89 9 49,3	22,1	25	92 32 32,5	46,0
50	85 26 33,2	4,8	50	89 15 11,7	21,3	50	92 37 18,5	45,2
97,75	85 32 38,0		107,75	89 20 33,0		117,75	92 42 3,7	
		6 3,6			5 20,4			4 44,3
98,00	85 38 41,6	2,5	108,00	89 25 53,4	5 19,4	118,00	92 46 48,0	43,6
25	85 44 44,1	1,3	25	89 31 12,8	18,4	25	92 51 31,6	42,7
50	85 50 45,4	6 0,1	50	89 36 31,2	17,4	50	92 56 14,3	41,9
98,75	85 56 45,5		108,75	89 41 48,6		118,75	93 0 56,2	
		5 58,9			5 16,5			4 41,1
99,00	86 2 44,4	57,8	109,00	89 47 5,1	15,5	119,00	93 5 37,3	4 40,2
25	86 8 42,2	56,7	25	89 52 20,6	14,5	25	93 10 17,5	39,5
50	86 14 38,9	55,4	50	89 57 35,1	13,6	50	93 14 57,0	38,6
99,75	86 20 34,3		109,75	90 2 48,7		119,75	93 19 35,6	
		5 54,4			5 12,6			4 38,0

TABLE III. *continued.*

Days.	D. M. S.	M. S.	Days.	D. M. S.	M. S.	Days.	D. M. S.	M. S.
120,00	93 24 13,6	4 37,1	130,00	96 19 14,0	4 8,0	140,00	98 56 22,5	43,4
25	93 28 50,7	36,3	25	96 23 22,0	7,4	25	99 0 5,9	42,9
50	93 33 27,0	35,5	50	96 27 29,4	6,7	50	99 3 48,8	42,3
120,75	93 38 2,5	4 34,7	130,75	96 31 36,1	4 6,1	140,75	99 7 31,1	3 41,8
121,00	93 42 37,2	33,9	131,00	96 35 42,2	5,3	141,00	99 11 12,9	3 41,2
25	93 47 11,1	33,2	25	96 39 47,5	4,8	25	99 14 54,1	40,6
50	93 51 44,3	32,4	50	96 43 52,3	4,1	50	99 18 34,7	40,1
121,75	93 56 16,7	4 31,7	131,75	96 47 56,4	4 3,4	141,75	99 22 14,8	3 39,5
122,00	94 0 48,4	30,8	132,00	96 51 59,8	2,8	142,00	99 25 54,3	39,0
25	94 5 19,2	30,1	25	96 56 2,6	2,1	25	99 29 33,3	38,4
50	94 9 49,3	29,4	50	97 0 4,7	1,5	50	99 33 11,7	37,9
122,75	94 14 18,7	4 28,1	132,75	97 4 6,2	4 0,9	142,75	99 36 49,6	3 37,3
123.00	94 18 47,3	27,8	133,00	97 8 7,1	4 0,2	143,00	99 40 26,9	36,8
25	94 23 15,1	27,1	25	97 12 7,3	3 59,6	25	99 44 3,7	36,3
50	94 27 42,2	26,4	50	97 16 6,9	59,0	50	99 47 40,0	35,7
123.75	94 32 8,6	4 25,6	133,75	97 20 5,9	58,3	143,75	99 51 15,7	3 35,2
124,00	94 36 34,2	24,8	134,00	97 24 4,2	57,7	144,00	99 54 50,9	34,7
25	94 40 59,0	24,1	25	97 28 1,9	57,1	25	99 58 25,6	34,1
50	94 45 23,1	23,4	50	97 31 59,0	56,5	50	100 1 59,7	33,6
124,75	94 49 46,5	4 22,7	134,75	97 35 55,5	3 55,8	144,75	100 5 33,3	3 33,1
125,00	94 54 9,2	22,0	135,00	97 39 51,3	55,3	145,00	100 9 6,4	32,5
25	94 58 31,2	21,2	25	97 43 46,6	54,6	25	100 12 38,9	32,0
50	95 2 52,4	20,5	50	97 47 41,2	54,0	50	100 16 10,9	31,5
125,75	95 7 12,9	4 19,8	135,75	97 51 35,2	3 53,4	145,75	100 19 42,4	3 31,0
126,00	95 11 32,7	19,0	136,00	97 55 28,6	52,8	146,00	100 23 13,4	30,5
25	95 15 51,7	18,4	25	97 59 21,4	52,2	25	100 26 43,9	30,0
50	95 20 10,1	17,6	50	98 3 13,6	51,6	50	100 30 13,9	29,4
126,75	95 24 27,7	4 17,0	136,75	98 7 5,2	3 51,0	146,75	100 33 43,3	3 28,9
127,00	95 28 44,7	16,2	137,00	98 10 56,2	50,4	147,00	100 37 12,2	28,4
25	95 33 0,9	15,5	25	98 14 46,6	49,8	25	100 40 40.6	27,9
50	95 37 16,4	14,9	50	98 18 36,4	49,3	50	100 44 8,5	27,4
127,75	95 41 31,3	4 14,1	137,75	98 22 25,7	3 48,6	147,75	100 47 35,9	3 26,9
128,00	95 45 45,4	13,5	138,00	98 26 14,3	48,1	148,00	100 51 2,8	26,4
25	95 49 58,9	12,8	25	98 30 2,4	47,4	25	100 54 29,2	25,9
50	95 54 11,7	12,0	50	98 33 49,8	46,9	50	100 57 55,1	25,4
128,75	95 58 23,7	4 11,5	138,75	98 37 36,7	3 46,3	148,75	101 1 20,5	3 24,9
129,00	96 2 35,2	10,7	139,00	98 41 23,0	45,8	149,00	101 4 45,4	24,5
25	96 6 45,9	10,0	25	98 45 8,8	45,1	25	101 8 9,9	23,9
50	96 10 55,9	9,4	50	98 48 53,9	44,6	50	101 11 33,8	23,4
129,75	96 15 5,3	4 8,7	139,75	98 52 38,5	3 44,0	149,75	101 14 57,2	22,9

TABLE III. *continued.*

Days.	D.	M.	S.	M.	S.	Days.	D.	M.	S.	M.	S.	Days.	D.	M.	S.	M.	S.
150,00	101	18	20,1			160,00	103	27	18,5	3	4,5	170,00	105	25	4,7		
25	101	21	42,6	3	22,5	25	103	30	23,0		4,0	25	105	27	53,5	2	48,8
50	101	25	4,6		22,0	50	103	33	27,0		3,6	50	105	30	42,0		48,5
150,75	101	28	26,1		21,5	160,75	103	36	30,6			170,75	105	33	30,1		48,1
				3	21,0					3	3,2					2	47,8
151,00	101	31	47,1		20,6	161,00	103	39	33,8		2,8	171,00	105	36	17,9		47,4
25	101	35	7,7		20,1	25	103	42	36,6		2,3	25	105	39	5,3		47,0
50	101	38	27,8		19,5	50	103	45	38,9		2,0	50	105	41	52,3		46,7
151,75	101	41	47,3			161,75	103	48	40,9			171,75	105	44	39,0		
				3	19,2					3	1,5					2	46,4
152,00	101	45	6,5	3	18,6	162,00	103	51	42,4		1,2	172,00	105	47	25,4		46,0
25	101	48	25,1		18,2	25	103	54	43,6		0,7	25	105	50	11,4		45,6
50	101	51	43,3		17,7	50	103	57	44,3		0,3	50	105	52	57,0		45,3
152,75	101	55	1,0			162,75	104	0	44,6			172,75	105	55	42,3		
				3	17,2					3	0,0					2	44,9
153,00	101	58	18,2		16,8	163,00	104	3	44,6	2	59,5	173,00	105	58	27,2		44,6
25	102	1	35,0		16,3	25	104	6	44,1		59,1	25	106	1	11,8		44,2
50	102	4	51,3		15,9	50	104	9	43,2		58,8	50	106	3	56,0		43,9
153,75	102	8	7,2			163,75	104	12	42,0			173,75	106	6	39,9		
				3	15,4					2	58,3					2	43,5
154,00	102	11	22,6		14,9	164,00	104	15	40,3		57,9	174,00	106	9	23,4		43,2
25	102	14	37,5		14,5	25	104	18	38,2		57,6	25	106	12	6,6		42,9
50	102	17	52,0		14,0	50	104	21	35,8		57,1	50	106	14	49,5		42,5
154,75	102	21	6,0			164,75	104	24	32,9			174,75	106	17	32,0		
				3	13,6					2	56,8					2	42,2
155,00	102	24	19,6		13,1	165,00	104	27	29,7		56,3	175,00	106	20	14,2		41,8
25	102	27	32,7		12,7	25	104	30	26,0		56,0	25	106	22	56,0		41,5
50	102	30	45,4		12,2	50	104	33	22,0		55,6	50	106	25	37,5		41,1
155,75	102	33	57,6			165,75	104	36	17,6			175,75	106	28	18,6		
				3	11,8					2	55,2					2	40,8
156,00	102	37	9,4		11,3	166,00	104	39	12,8		54,8	176,00	106	30	59,4		40,5
25	102	40	20,7		10,9	25	104	42	7,6		54,5	25	106	33	39,9		40,2
50	102	43	31,6		10,5	50	104	45	2,1		54,0	50	106	36	20,1		39,8
156,75	102	46	42,1			166,75	104	47	56,1			176,75	106	38	59,9		
				3	10,0					2	53,7					2	39,5
157,00	102	49	52,1		9,6	167,00	104	50	49,8		53,3	177,00	106	41	39,4		39,2
25	102	53	1,7		9,1	25	104	53	43,1		52,9	25	106	44	18,6		38,8
50	102	56	10,8		8,7	50	104	56	36,0		52,6	50	106	46	57,4		38,5
157,75	102	59	19,5			167,75	104	59	28,6			177,75	106	49	35,9		
				3	8,3					2	52,1					2	38,2
158,00	103	2	27,8		7,8	168,00	105	2	20,7		51,8	178,00	106	52	14,1		37,8
25	103	5	35,6		7,4	25	105	5	12,5		51,4	25	106	54	51,9		37,5
50	103	8	43,0		7,0	50	105	8	3,9		51,1	50	106	57	29,4		37,2
158,75	103	11	50,0			168,75	105	10	55,0			178,75	107	0	6,6		
				3	6,6					2	50,6					2	36,9
159,00	103	14	56,6		6,1	169,00	105	13	45,6		50,3	179,00	107	2	43,5		36,5
25	103	18	2,7		5,7	25	105	16	35,9		50,0	25	107	5	20,0		36,2
50	103	21	8,4		5,3	50	105	19	25,9		49,6	50	107	7	56,2		35,9
159,75	103	24	13,7			169,75	105	22	15,5			179,75	107	10	32,1		
				3	4,8					2	49,2					2	35,6

TABLE III. *continued.*

Days.	D. M. S.	M. S.	Days.	D. M. S.	M. S.	Days.	D. M. S.	M. S.
180,00	107 13 7,7		190,00	108 52 41,3		200,00	110 24 47,1	
25	107 15 43,0	2 35,3	25	108 55 4,6	2 23,3	5	110 29 12,6	4 25,5
50	107 18 18,0	35,0	50	108 57 27,7	23,1	201,0	110 33 37,1	24,5
180,75	107 20 52,6	34,6	190,75	108 59 50,5	22,8	5	110 38 0,7	23,6
181,00	107 23 26,9	2 34,3	191,00	109 2 13,1	2 22,6	202,0	110 42 23,2	4 22,5
25	107 26 0,9	34,0	25	109 4 35,3	22,2	5	110 46 44,8	21,6
50	107 28 34,6	33,7	50	109 6 57,3	22,0	203,0	110 51 5,5	20,7
181,75	107 31 8,0	33,4	191,75	109 9 19,0	21,7	5	110 55 25,2	19,7
182,00	107 33 41,1	2 33,1	192,00	109 11 40,4	2 21,4	204,0	110 59 43,9	4 18,7
25	107 36 13,9	32,8	25	109 14 1,6	21,2	5	111 4 1,7	17,8
50	107 38 46,3	32,4	50	109 16 22,5	20,9	205,0	111 8 18,6	16,9
182,75	107 41 18,4	32,1	192,75	109 18 43,1	20,6	5	111 12 34,6	16,0
183,00	107 43 50,2	2 31,8	193,00	109 21 3,5	2 20,4	206,0	111 16 49,6	4 15,0
25	107 46 21,8	31,6	25	109 23 23,5	20,0	5	111 21 3,7	14,1
50	107 48 53,0	31,2	50	109 25 43,3	19,8	207,0	111 25 16,9	13,2
183,75	107 51 24,0	31,0	193,75	109 28 2,9	19,6	5	111 29 29,2	12,3
184,00	107 53 54,6	2 30,6	194,00	109 30 22,2	2 19,3	208,0	111 33 40,5	4 11,3
25	107 56 24,9	30,3	25	109 32 41,2	19,0	5	111 37 51,0	10,5
50	107 58 54,9	30,0	50	109 34 59,9	18,7	209,0	111 42 0,6	9,6
184,75	108 1 24,7	29,8	194,75	109 37 18,4	18,5	5	111 46 9,3	8,7
185,00	108 3 54,1	2 29,4	195,00	109 39 36,7	2 18,3	210,0	111 50 17,1	4 7,8
25	108 6 23,2	29,1	25	109 41 54,6	17,9	5	111 54 24,1	7,0
50	108 8 52,0	28,8	50	109 44 12,3	17,7	211,0	111 58 30,2	6,1
185,75	108 11 20,5	28,5	195,75	109 46 29,8	17,5	5	112 2 35,4	5,2
186,00	108 13 48,8	2 28,3	196,00	109 48 46,9	2 17,1	212,0	112 6 39,7	4 4,3
25	108 16 16,7	27,9	25	109 51 3,9	17,0	5	112 10 43,2	3,5
50	108 18 44,4	27,7	50	109 53 20,6	16,7	213,0	112 14 45,8	2,6
186,75	108 21 11,7	27,3	196,75	109 55 36,9	16,3	5	112 18 47,6	1,8
187,00	108 23 38,8	2 27,1	197,00	109 57 53,1	2 16,2	214,0	112 22 48,5	4 0,9
25	108 26 5,6	26,8	25	110 0 9,0	15,9	5	112 26 48,6	4 0,1
50	108 28 32,1	26,5	50	110 2 24,6	15,6	215,0	112 30 47,9	3 59,3
187,75	108 30 58,3	26,2	197,75	110 4 40,0	15,4	5	112 34 46,4	58,5
188,00	108 33 24,2	2 25,9	198,00	110 6 55,1	2 15,1	216,0	112 38 44,0	3 57,6
25	108 35 49,8	25,6	25	110 9 10,0	14,9	5	112 42 40,8	56,8
50	108 38 15,1	25,3	50	110 11 24,6	14,6	217,0	112 46 36,8	56,0
188,75	108 40 40,2	25,1	198,75	110 13 39,0	14,4	5	112 50 31,9	55,1
189,00	108 43 5,0	2 24,8	199,00	110 15 53,1	2 14,1	218,0	112 54 26,3	3 54,4
25	108 45 29,5	24,5	.25	110 18 7,0	13,9	5	112 58 19,9	53,6
50	108 47 53,7	24,2	50	110 20 20,6	13,6	219,0	113 2 12,7	52,8
189,75	108 50 17,6	23,9	199,75	110 22 34,0	13,4	5	113 6 4,7	52,0
		2 23,7			2 13,1			3 51,2

3 I

TABLE III. *continued.*

Days.	D. M. S.	M. S.	Days.	D. M. S.	M. S.	Days.	D. M. S.	M. S.
220,0	113 9 55,9	3 50,4	240,0	115 34 5,1	3 22,4	260,0	117 41 18,3	2 59,5
5	113 13 46,3	49,7	5	115 37 27,5	21,7	5	117 44 17,8	58,9
221,0	113 17 36,0	48 8	241,0	115 40 49,2	21,1	261,0	117 47 16,7	58,5
5	113 21 24,8		5	115 44 10,3		5	117 50 15,2	
		3 48,1			3 20,4			2 58,0
222,0	113 25 12,9	47,?	242,0	115 47 30,7	19,9	262,0	117 53 13,2	57,4
5	113 29 0,2	46,6	5	115 50 50,6	19,2	5	117 56 10,6	57,?
223,0	113 32 46,8	45,9	243,0	115 54 9,8	18,6	263,0	117 59 7,6	56,4
5	113 36 32,7		5	115 57 28,4		5	118 2 4,0	
		3 45,0			3 18,0			2 56,0
224,0	113 40 17,7	44,3	244,0	116 0 46,4	17,4	264,0	118 5 0,0	55,4
5	113 44 2,0	43,5	5	116 4 3,8	16,8	5	118 7 55,4	55,0
225,0	113 47 45,5	42,8	245,0	116 7 20,6	16,2	265,0	118 10 50,4	54,4
5	113 51 28,3		5	116 10 36,8		5	118 13 44,8	
		3 42,1			3 15,6			2 54,0
226,0	113 55 10,4	41,4	246,0	116 13 52,4	15,0	266,0	118 16 38,8	53,5
5	113 58 51,8	40,6	5	116 17 7,4	14,4	5	118 19 32,3	53,0
227,0	114 2 32,4	39,9	247,0	116 20 21,8	13,8	267,0	118 22 25,3	52,5
5	114 6 12,3		5	116 23 35,6		5	118 25 17,8	
		3 39,2			3 13,3			2 52,0
228,0	114 9 51,5	38,5	248,0	116 26 48,9	12,6	268,0	118 28 9,8	51,5
5	114 13 30,0	37,7	5	116 30 1,5	12,1	5	118 31 1,3	51,1
229,0	114 17 7,7	37,0	249,0	116 33 13,6	11,5	269,0	118 33 52,4	50,5
5	114 20 44,7		5	116 36 25,1		5	118 36 42,9	
		3 36,3			3 10,9			2 50,1
230,0	114 24 21,0	35,7	250,0	116 39 36,0	10,4	270,0	118 39 33,0	49,7
5	114 27 56,7	34,9	5	116 42 46,4	9,8	5	118 42 22,7	49,1
231,0	114 31 31,6	34,3	251,0	116 45 56,2	9,2	271,0	118 45 11,8	48,7
5	114 35 5,9		5	116 49 5,4		5	118 48 0,5	
		3 33,5			3 8,6			248,2
232,0	114 38 39,4	32,9	252,0	116 52 14,0	8,1	272,0	118 50 48,7	47,8
5	114 42 12,3	32,1	5	116 55 22,1	7,6	5	118 53 36,5	47,3
233,0	114 45 44,4	31,5	253,0	116 58 29,7	7,0	273,0	118 56 23,8	46,9
5	114 48 15,9		5	117 1 36,7		5	118 59 10,7	
		3 30,8			3 6,4			2 46,3
234,0	114 52 46,7	30,2	254,0	117 4 43,1	5,9	274,0	119 1 57,0	46,0
5	114 56 16,9	29,5	5	117 7 49,0	5 3	5	119 4 43,0	45,5
235,0	114 59 46,4	28,8	255,0	117 10 54,3	4,8	275,0	119 7 28,5	45,0
5	115 3 15,2		5	117 13 59,1		5	119 10 13,5	
		3 28,1			3 4,3			2 44,6
236,0	115 6 43,3	27,5	256,0	117 17 3,4	3,7	276,0	119 12 58,1	44,1
5	115 10 10,8	26,9	5	117 20 7,1	3,2	5	119 15 42,2	43,7
237,0	115 13 37,7	26,2	257,0	117 23 10,3	2,6	277,0	119 18 25,9	43,2
5	115 17 3,9		5	117 26 12,9		5	119 21 9,1	
		3 25,5			3 2,1			2 42,8
238,0	115 20 29,4	24,9	258,0	117 29 15,0	1,6	278,0	119 23 51,9	42,4
5	115 23 54,3	24,2	5	117 32 16,6	1,1	5	119 26 34,3	41,9
239,0	115 27 18,5	23,6	259,0	117 35 17,7	0,5	279,0	119 29 16,2	41,6
5	115 30 42,1		5	117 38 18,2		5	119 31 57,7	
		3 23,0			3 0,1			2 41,1

TABLE III. *continued.*

Days.	D.	M.	S.	M.	S.	Days.	D.	M.	S.	M.	S.	Days.	D.	M.	S.	M.	S.
280,0	119	34	38,8			300,0	121	16	27,5			320,0	122	48	34,3		
5	119	37	19,5	2	40,7	5	121	18	52,3	2	24,8	5	122	50	45,8	2	11,5
281,0	119	39	59,7		40,2	301,0	121	21	16,8		24,5	321,0	122	52	57,0		11,2
5	119	42	39,4		39,7	5	121	23	41,0		24,2	5	122	55	7,9		10,9
				2	39,1					2	23,7					2	10,5
282,0	119	45	18,8			302,0	121	26	4,7			322,0	122	57	18,4		
5	119	47	57,7		38,9	5	121	28	28,2		23,5	5	122	59	28,7		10,3
283,0	119	50	36,2		38,5	303,0	121	30	51,2		23,0	323,0	123	1	38,7		10,0
5	119	53	14,3		38,1	5	121	33	13,9		22,7	5	123	3	48,4		9,7
				2	37,7					2	22,5					2	9,4
284,0	119	55	52,0			304,0	121	35	36,4			324,0	123	5	57,8		
5	119	58	29,2		37,2	5	121	37	58,4		22,0	5	123	8	6,9		9,1
285,0	120	1	6,1		36,9	305,0	121	40	20,0		21,6	325,0	123	10	15,6		8,7
5	120	3	42,5		36,1	5	121	42	41,3		21,3	5	123	12	24,1		8,5
				2	36,0					2	20,9					2	8,2
286,0	120	6	18,5			306,0	121	45	2,2			326,0	123	14	32,3		
5	120	8	54,1		35,6	5	121	47	22,8		20,6	5	123	16	40,2		7,9
287,0	120	11	29,4		35,3	307,0	121	49	43,1		20,3	327,0	123	18	47,8		7,6
5	120	14	4,1		34,7	5	121	52	3,1		20,0	5	123	20	55,1		7,3
				2	34,4					2	19,6					2	7,0
288,0	120	16	38,5			308,0	121	54	22,7			328,0	123	23	2,1		
5	120	19	12,5		34,0	5	121	56	41,9		19,2	5	123	25	8,9		6,8
289,0	120	21	46,1		33,6	309,0	121	59	0,8		18,9	329,0	123	27	15,3		6,4
5	120	24	19,3		33,2	5	122	1	19,4		18,6	5	123	29	21,5		6,2
				2	32,8					2	18,3					2	5,8
290,0	120	26	52,1			310,0	122	3	37,7			330,0	123	31	27,3		
5	120	29	24,5		32,4	5	122	5	55,6		17,9	5	123	33	32,9		5,6
291,0	120	31	56,6		32,1	311,0	122	8	13,1		17,5	331,0	123	35	38,2		5,3
5	120	34	28,2		31,6	5	122	10	30,4		17,3	5	123	37	43,2		5,0
				2	31,2					2	16,9					2	4,8
292,0	120	36	59,4			312,0	122	12	47,3			332,0	123	39	48,0		
5	120	39	30,8		30,9	5	122	15	3,9		16,6	5	123	41	52,4		4,4
293,0	120	42	0,7		30,4	313,0	122	17	20,2		16,3	333,0	123	43	56,6		4,2
5	120	44	30,8		30,1	5	122	19	36,1		15,9	5	123	46	0,5		3,9
				2	29,7					2	15,6					2	3,6
294,0	120	47	0,5			314,0	122	21	51,7			334,0	123	48	4,1		
5	120	49	29,8		29,3	5	122	24	7,0		15,3	5	123	50	7,4		3,3
295,0	120	51	58,7		28,9	315,0	122	26	22,0		15,0	335,0	123	52	10,5		3,1
5	120	54	27,3		28,6	5	122	28	36,6		14,6	5	123	54	13,3		2,8
				2	28,2					2	14,3					2	2,5
296,0	120	56	55,5			316,0	122	30	50,9			336,0	123	56	15,8		
5	120	59	23,3		27,8	5	122	33	4,9		14,0	5	123	58	18 0		2,2
297,0	121	1	50,7		27,4	317,0	122	35	18,6		13,7	337,0	124	0	20,0		2,0
5	121	4	17,7		27,0	5	122	37	32,0		13,4	5	124	2	21,7		1,7
				2	26,7					2	13,1					2	1,4
298,0	121	6	44,4			318,0	122	39	45,1			338,0	124	4	23,1		
5	121	9	10,7		26,3	5	122	41	57,8		12,7	5	124	6	24,3		1,2
299,0	121	11	36,7		26,0	319,0	122	44	10,3		12,5	339,0	124	8	25,2		0,9
5	121	14	2,3		25,6	5	122	46	22,4		12,1	5	124	10	25,8		0,6
				2	25,2					2	11,9					2	0,3

TABLE III. *continued.*

Days.	D.	M.	S.	M.	S.	Days.	D.	M.	S.	M.	S.	Days.	D.	M.	S.	M.	S.
340,0	124	12	26,1			360,0	125	29	13,3	1	50,3	380,0	126	39	53,0	1	41,7
5	124	14	26,2	2	0,1	5	125	31	3,6		50,0	5	126	41	34,7		41,5
341,0	124	16	26,0	1	59,8	361,0	125	32	53,6		49,8	381,0	126	43	16,2		41,3
5	124	18	25,6		59,6	5	125	34	43,4	1	49,6	5	126	44	57,5	1	41,1
					59,3												
342,0	124	20	24,9		59,0	362,0	125	36	33,0		49,3	382,0	126	46	38,6		40,9
5	124	22	23,9		58,8	5	125	38	22,3		49,2	5	126	48	19,5		40,7
343,0	124	24	22,7		58,5	363,0	125	40	11,5		48,9	383,0	126	50	0,2		40,5
5	124	26	21,2	1	58,3	5	125	42	0,4	1	48,6	5	126	51	40,7	1	40,4
344,0	124	28	19,5		58,0	364,0	125	43	49,0		48,5	384,0	126	53	21,1		40,1
5	124	30	17,5		57,8	5	125	45	37,5		48,2	5	126	55	1,2		39,9
345,0	124	32	15,3		57,4	365,0	125	47	25,7		48,1	385,0	126	56	41,1		39,8
5	124	34	12,7	1	57,3	5	125	49	13,8	1	47,8	5	126	58	20,9	1	39,5
346,0	124	36	10,0		57,0	366,0	125	51	1,6		47,5	386,0	127	0	0,4		39,4
5	124	38	7,0		56,7	5	125	52	49,1		47,4	5	127	1	39,8		39,1
347,0	124	40	3,7		56,5	367,0	125	54	36,5		47,1	387,0	127	3	18,9		39,0
5	124	42	0,2	1	56,2	5	125	56	23,6	1	46,9	5	127	4	57,9	1	38,8
348,0	124	43	56,4		56,0	368,0	125	58	10,5		46,7	388,0	127	6	36,7		38,6
5	124	45	52,4		55,7	5	125	59	57,2		46,5	5	127	8	15,3		38,4
349,0	124	47	48,1		55,5	369,0	126	1	43,7		46,3	389,0	127	9	53,7		38,2
5	124	49	43,6	1	55,3	5	126	3	30 0	1	46,0	5	127	11	31,9	1	38,0
350,0	124	51	38,9		55,0	370,0	126	5	16,0		45,8	390,0	127	13	9,9		37,8
5	124	53	33,9		54,8	5	126	7	1,8		45,7	5	127	14	47,7		37,7
351,0	124	55	28,7		54,5	371,0	126	8	47,5		45,4	391,0	127	16	25,4		37,4
5	124	57	23,2	1	54,2	5	126	10	32,9	1	45,1	5	127	18	2,8	1	37,3
352,0	124	59	17,4		54,1	372,0	126	12	18,0		45,0	392,0	127	19	40,1		37,1
5	125	1	11,5		53,8	5	126	14	3,0		44,8	5	127	21	17,2		36,9
353,0	125	3	5,3		53,5	373,0	126	15	47,8		44,6	393,0	127	22	54,1		36,7
5	125	4	58,8	1	53,3	5	126	17	32,4	1	44,4	5	127	24	30,8	1	36,6
354,0	125	6	52,1		53,1	374,0	126	19	16,8		44,1	394,0	127	26	7,4		36,3
5	125	8	45,2		52,8	5	126	21	0,9		43,9	5	127	27	43,7		36,2
355,0	125	10	38,0		52,6	375,0	126	22	44,8		43,8	395,0	127	29	19,9		36,0
5	125	12	30,6	1	52,3	5	126	24	28,6	1	43,5	5	127	30	55,9	1	35,8
356,0	125	14	22,9		52,1	376,0	126	26	12,1		43,3	396,0	127	32	31,7		35,7
5	125	16	15,0		51,9	5	126	27	55,4		43,1	5	127	34	7,4		35,4
357,0	125	18	6,9		51,7	377,0	126	29	38,5		43,0	397,0	127	35	42,8		35,3
5	125	19	58,6	1	51,4	5	126	31	21,5	1	42,7	5	127	37	18,1	1	35,1
358,0	125	21	50,0		51,2	378,0	126	33	4,2		42,5	398,0	127	38	53,2		34,9
5	125	23	41,2		50,9	5	126	34	46,7		42,3	5	127	40	28,1		34,7
359,0	125	25	32,1		50,7	379,0	126	36	29,0		42,1	399,0	127	42	2,8		34,6
5	125	27	22,8	1	50,5	5	126	38	11,1	1	41,9	5	127	43	37,4	1	34,4

TABLE III. *continued.*

Days.	D.	M.	S.	M. S.	Days.	D.	M.	S.	M. S.	Days.	D.	M.	S.	M. S.
400	127	45	11,8		440	129	42	16,5		480	131	24	31,0	
401	127	48	20,0	3 8,2	441	129	44	59,9	2 43,4	481	131	26	54,6	2 23,6
402	127	51	27,6	7,6	442	129	47	42,7	42,8	482	131	29	17,8	23,2
403	127	54	34,4	6,8	443	129	50	25,0	42,3	483	131	31	40,5	22,7
404	127	57	40,6	3 6,2	444	129	53	6,7	2 41,7	484	131	34	2,7	2 22,2
405	128	0	46,1	5,5	445	129	55	47,9	41,2	485	131	36	24,5	21,8
406	128	3	50,8	4,7	446	129	58	28,5	40,6	486	131	38	45,9	21,4
407	128	6	55,0	4,2	447	130	1	8,6	40,1	487	131	41	6,9	21,0
408	128	9	58,4	3 3,4	448	130	3	48,2	2 39,6	488	131	43	27,4	2 20,5
409	128	13	1,2	2,8	449	130	6	27,3	39,1	489	131	45	47,5	20,1
410	128	16	3,3	2,1	450	130	9	5,8	38,5	490	131	48	7,2	19,7
411	128	19	4,7	1,4	451	130	11	43,8	38,0	491	131	50	26,5	19,3
412	128	22	5,5	3 0,8	452	130	14	21,3	2 37,5	492	131	52	45,3	2 18,8
413	128	25	5,7	3 0,2	453	130	16	58,3	37,0	493	131	55	3,8	18,5
414	128	28	5,2	2 59,5	454	130	19	34,7	36,4	494	131	57	21,8	18,0
415	128	31	4,0	58,8	455	130	22	10,7	36,0	495	131	59	39,3	17,5
416	128	34	2,2	2 58,2	456	130	24	46,1	2 35,4	496	132	1	56,5	2 17,2
417	128	36	59,8	57,6	457	130	27	21,1	35,0	497	132	4	13,3	16,8
418	128	39	56,7	56,9	458	130	29	55,5	34,4	498	132	6	29,7	16,4
419	128	42	53,0	56,3	459	130	32	29,4	33,9	499	132	8	45,7	16,0
420	128	45	48,7	2 55,7	460	130	35	2,9	2 33,5	500	132	11	1,2	2 15,5
421	128	48	43,8	55,1	461	130	37	35,8	32,9	501	132	13	16,4	15,2
422	128	51	38,2	54,4	462	130	40	8,3	32,5	502	132	15	31,1	14,7
423	128	54	32,0	53,8	463	130	42	40,2	31,9	503	132	17	45,5	14,4
424	128	57	25,3	2 53,3	464	130	45	11,7	2 31,5	504	132	19	59,5	2 14,0
425	129	0	17,9	52,6	465	130	47	42,7	31,0	505	132	22	13,1	13,6
426	129	3	9,9	52,0	466	130	50	13,2	30,5	506	132	24	26,3	13,2
427	129	6	1,3	51,4	467	130	52	43,2	30,0	507	132	26	39,1	12,8
428	129	8	52,1	2 50,8	468	130	55	12,8	2 29,6	508	132	28	51,5	2 12,4
429	129	11	42,3	50,2	469	130	57	41,8	29,0	509	132	31	3,5	12,0
430	129	14	31,9	49,6	470	131	0	10,4	28,6	510	132	33	15,2	11,7
431	129	17	20,9	49,0	471	131	2	38,5	28,1	511	132	35	26,4	11,2
432	129	20	9,4	2 48,5	472	131	5	6,2	2 27,7	512	132	37	37,3	2 10,9
433	129	22	57,3	47,9	473	131	7	33,4	27,2	513	132	39	47,8	10,5
434	129	25	44,6	47,3	474	131	10	0,1	26,7	514	132	41	58,0	10,2
435	129	28	31,3	46,7	475	131	12	26,4	26,3	515	132	44	7,7	9,7
436	129	31	17,5	2 46,2	476	131	14	52,2	2 25,8	516	132	46	17,1	2 9,4
437	129	34	3,1	45,6	477	131	17	17,6	25,4	517	132	48	26,1	9,0
438	129	36	48,1	45,0	478	131	19	42,5	24,9	518	132	50	34,8	8,7
439	129	39	32,6	44,5	479	131	22	7,0	24,5	519	132	52	43,0	8,2
				2 43,9					2 24 0					2 7,9

3 K

TABLE III. *continued.*

Days.	D. M. S.	M. S	Days.	D. M. S.	M. S.	Days.	D. M. S.	M S.
520	132 54 50,9		560	134 15 27,4		600	135 27 59,9	
521	132 56 58,5	2 7,6	561	134 17 21,7	1 54,3	601	135 29 43,2	1 43,3
522	132 59 5,7	7,2	562	134 19 15,8	54,1	602	135 31 26,3	43,1
523	133 1 12,5	6,8	563	134 21 9,5	53,7	603	135 33 9,1	42,8
		2 6,5			1 53,4			1 42,6
524	133 3 19,0		564	134 23 2,9		604	135 34 51,7	
525	133 5 25,1	6,1	565	134 24 56,1	53,2	605	135 36 34,0	42,3
526	133 7 30,9	5,8	566	134 26 49,0	52,9	606	135 38 16,0	42,0
527	133 9 36,3	5,4	567	134 28 41,5	52,5	607	135 39 57,8	41,8
		2 5,0			1 52,3			1 41,6
528	133 11 41,3		568	134 30 33,8		608	135 41 39,4	
529	133 13 46,0	4,7	569	134 32 25,8	52,0	609	135 43 20,8	41,4
530	133 15 50,4	4,4	570	134 34 17,5	51,7	610	135 45 1,9	41,1
531	133 17 54,4	4,0	571	134 36 8,9	51,4	611	135 46 42,7	40,8
		2 3,7			1 51,2			1 40,6
532	133 19 58,1		572	134 38 0,1		612	135 48 23,3	
533	133 22 1,4	3,3	573	134 39 50,9	50,8	613	135 50 3,6	40,3
534	133 24 4,4	3,0	574	134 41 41,4	50,5	614	135 51 43,7	40,1
535	133 26 7,1	2,7	575	134 43 31,7	50,3	615	135 53 23,6	39,9
		2 2,3			1 50,0			1 39,6
536	133 28 9,4		576	134 45 21,7		616	135 55 3,2	
537	133 30 11,3	1,9	577	134 47 11,4	49,7	617	135 56 42,6	39,4
538	133 32 13,0	1,7	578	134 49 0,9	49,5	618	135 58 21,8	39,2
539	133 34 14,3	1,3	579	134 50 50,0	49,1	619	136 0 0,7	38,9
		2 1,0			1 48,8			1 38,7
540	133 36 15,3		580	134 52 38,8		620	136 1 39,4	
541	133 38 15,9	0,6	581	134 54 27,4	48,6	621	136 3 17,9	38,5
542	133 40 16,2	0,3	582	134 56 15,7	48,3	622	136 4 56,1	38,2
543	133 42 16,2	2 0,0	583	134 58 3,8	48,1	623	136 6 34,1	38,0
		1 59,7			1 47,8			1 37,8
544	133 44 15,9		584	134 59 51,6		624	136 8 11,9	
545	133 46 15,2	59,3	585	135 1 39,1	47,5	625	136 9 49,4	37,5
546	133 48 14,2	59,0	586	135 3 26,4	47,3	626	136 11 26,7	37,3
547	133 50 12,9	58,7	587	135 5 13,4	47,0	627	136 13 3,8	37,1
		1 58,4			1 46,7			1 36,9
548	133 52 11,3		588	135 7 0,1		628	136 14 40,7	
549	133 54 9,4	58,1	589	135 8 46,5	46,4	629	136 16 17,3	36,6
550	133 56 7,1	57,7	590	135 10 32,6	46,1	630	136 17 53,7	36,4
551	133 58 4,6	57,5	591	135 12 18,5	45,9	631	136 19 29,9	36,2
		1 57,1			1 45,6			1 36,0
552	134 0 1,7		592	135 14 4,1		632	136 21 5,9	
553	134 1 58,5	56,8	593	135 15 49,5	45,4	633	136 22 41,6	35,7
554	134 3 54,9	56,4	594	135 17 34,6	45,1	634	136 24 17,1	35,5
555	134 5 51,1	56,2	595	135 19 19,5	44,9	635	136 25 52,4	35,3
		1 55,9			1 44,6			1 35,1
556	134 7 47,0		596	135 21 4,1		636	136 27 27,5	
557	134 9 42,5	55,5	597	135 22 48,4	44,3	637	136 29 2,3	34,8
558	134 11 37,8	55,3	598	135 24 32,5	44,1	638	136 30 37,0	34,7
559	134 13 32,7	54,9	599	135 26 16,3	43,8	639	136 32 11,4	34,4
		1 54,7			1 43,6			1 34,2

TABLE III. *continued.*

Days.	D.	M.	S.	M.	S.	Days.	D.	M.	S.	M.	S.	Days.	D.	M.	S.	M.	S.
640	136	33	45,6			680	137	33	45,5			750,0	139	7	14,7		
641	136	35	19,6	1	34,0	681	137	35	11,5	1	26,0	752,5	139	10	20,9	3	6,2
642	136	36	53,1		33,8	682	137	36	37,3		25,8	755,0	139	13	26,1		5,2
643	136	38	26,9		33,5	683	137	38	2,9		25,6	757,5	139	16	30.4		4,3
				1	33,4					1	25,5					3	3,5
644	136	40	0,3		33,1	684	137	39	28,4		25,3	760,0	139	19	33,9		2,5
645	136	41	33,4		32,9	685	137	40	53,7		25,1	762,5	139	22	36,4		1,7
646	136	43	6,3		32,7	686	137	42	18,8		24,9	765,0	139	25	38,1		0,9
647	136	44	39,0	1	32,5	687	137	43	43,7	1	24,7	767,5	139	28	39,0	3	0,0
648	136	46	11,5		32,	688	137	45	8,4		24,6	770,0	139	31	39,0	2	59,1
649	136	47	43,8		32,1	689	137	46	33,0		24,3	772,5	139	34	38,1		58,3
650	136	49	15,9		31,9	690	137	47	57,3		24,2	775,0	139	37	36,4		57,5
651	136	50	47,8	1	31,7	691	137	49	21,5	1	24,1	777,5	139	40	33,9	2	56,6
652	136	52	19,5		31,4	692	137	50	45,6		23,8	780,0	139	43	30,5		55,9
653	136	53	50,9		31,3	693	137	52	9,4		23,7	782,5	139	46	26,4		55,0
654	136	55	22,2		31,0	694	137	53	33,1		23,5	785,0	139	49	21,4		54,2
655	136	56	53,2	1	30,9	695	137	54	56,6	1	23,3	787,5	139	52	15,6	2	53,4
656	136	58	24,1		30,6	696	137	56	19,9		23,1	790,0	139	55	9,0		52,6
657	136	59	54,7		30,5	697	137	57	43,0		23,0	782,5	139	58	1,6		51,8
658	137	1	25,2		30,2	698	137	59	6,0		22,8	785,0	140	0	53,4		51,0
659	137	2	55,4	1	30,1	699	138	0	28,8	1	22,6	797,5	140	3	44,4	2	50,3
660	137	4	25 5		29,8	700	138	1	51,4	3	25,8	800,0	140	6	34,7		49,4
661	137	5	55,3		29,7	702,5	138	5	17,2		24,7	802,5	140	9	24,1		48,7
662	137	7	25,0		29,4	705,0	138	8	41,9		23,7	805,0	140	12	12,8		48,0
663	137	8	54,4	1	29,3	707,5	138	12	5,6	3	22,6	807,5	140	15	0,8	2	47,2
664	137	10	23,7		29,0	710,0	138	15	28,2		21,6	810,0	140	17	48 0		46,4
665	137	11	52,7		28,9	712,5	138	18	49,8		20,5	822,5	140	20	34,4		45,7
666	137	13	21,6		28,6	715,0	138	22	10,3		19,5	825,0	140	23	20,1		45,0
667	137	14	50,2	1	28,5	717,5	138	25	29,8	3	18,6	817,5	140	26	5,1	2	44,3
668	137	16	18,7		28,	720,0	138	28	48,4		17,5	820,0	140	28	49,4		43,5
669	137	17	47,0		28,1	722,5	138	32	5,9		16,5	822,5	140	31	32,9		42,8
670	137	19	15,1		27,9	725,0	138	35	22,4		15,5	825,0	140	34	15,7		42,1
671	137	20	43,0	1	27,7	727,5	138	38	37,9	3	14,6	827,5	140	36	57,8	2	41,4
672	137	22	10,7		27,5	730,0	138	41	52,5		13,6	830,0	140	39	39,2		40,6
673	137	23	38,2		27,3	732,5	138	45	6,1		12,6	832,5	140	42	19,8		40,0
674	137	25	5,5		27,1	735,0	138	48	18,7		11,6	835,0	140	44	59,8		39,3
675	137	26	32,6	1	27,0	737,5	138	51	30,3	3	10,8	837,5	140	47	39,1	2	38,6
676	137	27	59,6		26,7	740,0	138	54	41,1		9,8	840,0	140	50	17,7		37,9
677	137	29	26,3		26,6	742,5	138	57	50,9		8,8	842,5	140	52	55,6		37,2
678	137	30	52,9		26,4	745,0	139	0	59,7		8,0	845,0	140	55	32,8		36,6
679	137	32	19,3	1	26,2	747,5	139	4	7,7	3	7,0	847,5	140	58	9,4	2	35,9

TABLE III. *continued.*

Days.	D.	M.	S.	M.	S.	Days.	D.	M.	S.	M.	S.	Days.	D.	M.	S.	M.	S.
850,0	141	0	45,3			950,0	142	36	24,4			1050,0	143	58	32,8		
852,5	141	3	20,5	2	35,2	952,5	142	38	36,6	2	12,2	1052,5	144	0	27,3	1	54,5
855,0	141	5	55,1		34,6	955,0	142	40	48,3		11,7	1055,0	144	2	21,4		51,1
857,5	141	8	29,0		33,9	957,5	142	42	59,6		11,3	1057,5	144	4	15,1		53,7
				2	33,3					2	10,7					1	53,4
860,0	141	11	2,3			960,0	142	45	10,3			1060,0	144	6	8,5		
862,5	141	13	34,9		32,6	962,5	142	47	20,5		10,2	1062,5	144	8	1,4		52,9
865,0	141	16	6,9		32,0	965,0	142	49	30,3		9,8	1065,0	144	9	54,0		52,6
867,5	141	18	38,3		31,4	967,5	142	51	39,6		9,3	1067,5	144	11	46,2		52,2
				2	30,7					2	8,7					1	51,8
870,0	141	21	9,0			970,0	142	53	48,3			1070,0	144	13	38,0		
872,5	141	23	39,1		30,1	972,5	142	55	56,6		8,3	1072,5	144	15	29,4		51,4
875,0	141	26	8,6		29,5	975,0	142	58	4,5		7,9	1075,0	144	17	20,5		51,1
877,5	141	28	37,4		28,8	977,5	143	0	11,8		7,3	1077,5	144	19	11,2		50,7
				2	28,3					2	6,9					1	50,3
880,0	141	31	5,7			980,0	143	2	18,7			1080,0	144	21	1,5		
882,5	141	33	33,3		27,6	982,5	143	4	25,2		6,5	1082,5	144	22	51,5		50,0
885,0	141	36	0,4		27,1	985,0	143	6	31,1		5,9	1085,0	144	24	41,0		49,5
887,5	141	38	26,8		26,1	987,5	143	8	36,6		5,5	1087,5	144	26	30,2		49,2
				2	25,9					2	5,1					1	48,9
890,0	141	40	52,7			990,0	143	10	41,7			1090,0	144	28	19,1		
892,5	141	43	17,9		25,2	992,5	143	12	46,3		4,6	1092,5	144	30	7,6		48,5
855,0	141	45	42,6		24,7	995,0	143	14	50,4		4,1	1095,0	144	31	55,8		48,2
897,5	141	48	6,6		24,0	997,5	143	16	54,1		3,7	1097,5	144	33	43,6		47,8
				2	23,6					2	3,3					1	47,4
900,0	141	50	30,2			1000,0	143	18	57,4			1100,0	144	35	31,0		
902,5	141	52	53,1		22,9	1002,5	143	21	0,2		2,8	1102,5	144	37	18,1		47,1
905,0	141	55	15,4		22,3	1005,0	143	23	2,5		2,3	1105,0	144	39	4,9		46,8
907,5	141	57	37,1		21,7	1007,5	143	25	4,5		2,0	1107,5	144	40	51,3		46,4
				2	21,3					2	1,5					1	46,1
910,0	141	59	58,4			1010,0	143	27	6,0			1110,0	144	42	37,4		
912,5	142	2	19,1		20,7	1002,5	143	29	7,0		1,0	1112,5	144	44	23,1		45,7
915,0	142	4	39,2		20,1	1005,0	143	31	7,6		0,6	1115,0	144	46	8,5		45,4
917,5	142	6	58,7		19,5	1017,5	143	33	7,8	2	0,2	1117,5	144	47	53,5		45,0
				2	19,0					1	59,8					1	44,6
920,0	142	9	17,7			1020,0	143	35	7,6			1120,0	144	49	38,1		
922,5	142	11	36,2		18,5	1022,5	143	37	7,0		59,4	1122,5	144	51	22,6		41,5
925,0	142	13	54,1		17,9	1025,0	143	39	5,9		58,9	1125,0	144	53	6,7		44,1
927,5	142	16	11,5		17,4	1027,5	143	41	4,4		58,5	1127,5	144	54	50,4		43,7
				2	16,9					1	58,1					1	43,4
930,0	142	18	28,4			1030,0	143	43	2,5			1130,0	144	56	33,8		
932,5	142	20	44,7		16,3	1032,5	143	45	0,2		57,7	1132,5	144	58	16,9		43,1
935,0	142	23	0,5		15,8	1035,0	143	46	57,5		57,3	1135,0	144	59	59,6		42,7
937,5	142	25	15,8		15,3	1037,5	143	48	54,4		56,9	1137,5	145	1	42,0		42,4
				2	14,8					1	56,5					1	42,1
940,0	142	27	30,6			1040,0	143	50	50,9			1140,0	145	3	24,1		
942,5	142	29	44,8		14,2	1042,5	143	52	47,0		56,1	1142,5	145	5	5,9		41,8
945,0	142	31	58	5	13,7	1045,0	143	54	42,6		55,6	1145,0	145	6	47,3		41,4
947,5	142	34	11,7		13,2	1047,5	143	56	37,9		55,3	1147,5	145	8	28,5		41,2
				2	12,7					1	54,9					1	40,8

TABLE III. *continued.*

Days.	D. M. S.	M. S.	Days.	D. M. S.	M. S.	Days.	D. M. S.	M. S.
1150,0	145 10 9,3		1300	146 42 19,6		1500	148 24 0,8	
1152,5	145 11 49,8	1 40,5	1305	146 45 8,1	2 48,5	1505	148 26 18,4	2 17,6
1155,0	145 13 30,0	40,2	1310	146 47 55,7	47,6	1510	148 28 35,3	16,9
1157,5	145 15 9,8	39,8	1315	146 50 42,4	46,7	1515	148 30 51,6	16,3
1160,0	145 16 49,4	1 39,6	1320	146 53 28,2	2 45,8	1520	148 33 7,3	2 15,7
1162,5	145 18 28,7	39,3	1325	146 56 13,1	44,9	1525	148 35 22,3	15,0
1165,0	145 20 7,7	39,0	1330	146 58 57,2	44,1	1530	148 37 36,7	14,4
1167,5	145 21 46,3	38,6	1335	147 1 40,3	43,1	1535	148 39 50,5	13,8
1170,0	145 23 24,7	1 38,4	1340	147 4 22,6	2 42,3	1540	148 42 3,6	2 13,1
1172,5	145 25 2,8	38,1	1345	147 7 4,0	41,4	1545	148 44 16,2	12,6
1175,0	145 26 40,5	37,7	1350	147 9 44,6	40,6	1550	148 46 28,1	11,9
1177,5	145 28 18,0	37,5	1355	147 12 24,3	39,7	1555	148 48 39,4	11,3
1180,0	145 29 55,1	1 37,1	1360	147 15 3,2	2 38,9	1560	148 50 50,2	2 10,8
1182,5	145 31 32,0	36,9	1365	147 17 41,3	38,1	1565	148 53 0,3	10,1
1185,0	145 33 8,6	36,6	1370	147 20 18,5	37,2	1570	148 55 9,9	9,6
1187,5	145 34 44,9	36,3	1375	147 22 55,0	36,5	1575	148 57 18,8	8,9
1190,0	145 36 20,9	1 36,0	1380	147 25 30,6	2 35,6	1580	148 59 27,3	2 8,5
1192,5	145 37 56,6	35,7	1385	147 28 5,4	34,8	1585	149 1 35,1	7,8
1195,0	145 39 32,0	35,4	1390	147 30 39,5	34,1	1590	149 3 42,3	7,2
1197,5	145 41 7,2	35,2	1395	147 33 12,7	33,2	1595	149 5 49,0	6,7
1200	145 42 42,1	1 34,9	1400	147 35 45,2	2 32,5	1600	149 7 55,2	2 6,2
1205	145 45 50,9	3 8,8	1405	147 38 16,9	31,7	1605	149 10 0,7	5,5
1210	145 48 58,7	7,8	1410	147 40 47,8	30,9	1610	149 12 5,7	5,0
1215	145 52 5,3	6,6	1415	147 43 17,9	30,1	1615	149 14 10,2	4,5
1220	145 55 10,9	3 5,6	1420	147 45 47,4	2 29,5	1620	149 16 14,2	2 4,0
1225	145 58 15,4	4,5	1425	147 48 16,0	28,6	1625	149 18 17,6	3,4
1230	146 1 18,8	3,4	1430	147 50 44,0	28,0	1630	149 20 20,4	2,8
1235	146 4 21,1	2,3	1435	147 53 11,2	27,2	1635	149 22 22,7	2,3
1240	146 7 22,4	3 1,3	1440	147 55 37,7	2 26,5	1640	149 24 24,5	2 1,8
1245	146 10 22,7	0,3	1445	147 58 3,4	25,7	1645	149 26 25,8	1,3
1250	146 13 21,9	2 59,2	1450	148 0 28,4	25,0	1650	149 28 26,5	0,7
1255	146 16 20,1	58,2	1455	148 2 52,8	24,4	1655	149 30 26,8	0,3
1260	146 19 17,3	2 57,2	1460	148 5 16,4	2 23,6	1660	149 32 26,5	1 59,7
1265	146 22 13,5	56,2	1465	148 7 39,3	22,9	1665	149 34 25,7	59,2
1270	146 25 8,7	55,2	1470	148 10 1,6	22,3	1670	149 36 24,4	58,7
1275	146 28 2,9	54,2	1475	148 12 23,1	21,5	1675	149 38 22,6	58,2
1280	146 30 56,1	2 53,2	1480	148 14 44,0	2 20,9	1680	149 40 20,3	1 57,7
1285	146 33 48,4	52,3	1485	148 17 4,2	20,2	1685	149 42 17,6	57,3
1290	146 36 39,8	51,4	1490	148 19 23,8	19,6	1690	149 44 14,3	56,7
1295	146 39 30,2	50,4	1495	148 21 42,6	18,8	1695	149 46 10,5	56,2
		2 49,4			2 18,2			1 55,8

3 L

TABLE III. *continued.*

Days.	D. M. S.	M. S.	Days.	D. M. S.	M. S.	Days.	D. M. S.	M. S.
1700	149 48 6,3		2000	151 31 2,0		2400	153 18 44,1	
1705	149 50 1,6	1 55,3	2010	151 34 5,1	3 3,1	2410	153 21 6,1	2 22,0
1710	149 51 56,4	54,8	2020	151 37 7,0	1,9	2420	153 23 27,3	21,2
1715	149 53 50,7	54,3	2030	151 40 7,6	0,6	2430	153 25 47,6	20,3
		1 53,9			2 59,3			2 19,5
1720	149 55 44,6		2040 -	151 43 6,9		2440	153 28 7,1	
1725	149 57 38,0	53,1	2050	151 46 5,1	58,2	2450	153 30 25,8	18,7
1730	149 59 30,9	52,9	2060	151 49 2,0	56,9	2460	153 32 43,7	17,9
1735	150 1 23,4	52,5	2070	151 51 57,8	55,8	2470	153 35 0,9	17,2
		1 52,0			2 54,5			2 16,4
1740	150 3 15,4		2080	151 54 52,3		2480 -	153 37 17,3	
1745	150 5 7,0	51,6	2090	151 57 45,7	53,4	2490	153 39 32,9	15,6
1750	150 6 58,1	51,1	2100	152 0 37,9	52,2	2500	153 41 47,8	14,9
1755	150 8 48,7	50,6	2110	152 3 28,9	51,0	2510	153 44 1,9	14,1
		1 50,2			2 49,9			2 13,4
1760	150 10 38,9		2120	152 6 18.8		2520	153 46 15,3	
1765	150 12 28.7	49,8	2130	152 9 7,7	48,2	2530	153 48 27,1	12,6
1770	150 14 18,1	49,2	2140	152 11 55,4	47,7	2540	153 50 39,8	11,9
1775	150 16 7,0	48,9	2150	152 14 42,0	46,6	2550	153 52 51,0	11,2
		1 48.5			2 45,5			2 10,4
1780	150 17 55,5		2160	152 17 27,5		2560	153 55 1,4	
1785	150 19 43,5	48,0	2170	152 20 11,9	44,4	2570	153 57 11,1	9,7
1790	150 21 31,2	47,7	2180	152 22 55,3	43,4	2580	153 59 20,2	9,1
1795	150 23 18,4	47,2	2190	152 25 37,7	42,4	2590	154 1 28,6	8,4
		1 46,8			2 41,3			2 7,7
1800	150 25 5,2		2200	152 28 19,0		2600	154 3 36,3	
1810	150 28 37,5	3 32,1	2210	152 30 59,3	40,3	2610	154 5 43,3	7,0
1820	150 32 8,1	30,6	2220	152 33 38,5	39,2	2620	154 7 49,6	6,3
1830	150 35 37,2	29,1	2230	152 36 16,8	38,3	2630	154 9 55,2	5,6
		3 27,4			2 37,3			2 5,0
1840	150 39 4,6		2240	152 38 54,1		2640	154 12 0,2	
1850	150 42 30,5	25,8	2250	152 41 30,4	36,1	2650	154 14 4,5	4,3
1860	150 45 54,8	24,3	2260	152 44 5,7	35,3	2660	154 16 8,2	3,7
1870	150 49 17,6	22,8	2270	152 46 40,1	31,4	2670	154 18 11,2	3,0
		3 21,2			2 33,4			2 2,4
1880	150 52 38,8		2280	152 49 13,5		2680	154 20 13,6	
1890	150 55 58,6	19,8	2290	152 51 46,0	32,5	2690	154 22 15,3	1,7
1900	150 59 16,8	18,2	2300	152 54 17,5	31,5	2700	154 24 16,4	1,1
1910	151 2 33,6	16,8	2310	152 56 48,1	30,6	2710	154 26 16,9	0,5
		3 15,4			2 29,7			1 59,9
1920.	151 5 49,0		2320	152 59 17,8		2720	154 28 16,8	
1930	151 9 2,9	13,9	2330	153 1 46,7	28,9	2730	154 30 16,1	59,3
1940	151 12 15,4	12,5	2340	153 4 14,7	28,0	2740	154 32 14,7	58,6
1950	151 15 26,5	11,1	2350	153 6 41,8	27,1	2750	154 34 12,8	58,1
		3 9,8			2 26,2			1 57,5
1960	151 18 36,3		2360	153 9 8,0		2760	154 36 10.3	
1970	151 21 44,7	8,4	2370	153 11 33,3	25,3	2770	154 38 7,1	56,8
1980	151 24 51,8	7,1	2380	153 13 57,7	24,4	2780	154 40 3,4	56,3
1990	151 27 57,5	5,7	2390	153 16 21,3	23,6	2790	154 41 59,1	55,7
		3 4,5			2 22,8			1 55,1

TABLE III. *continued.*

Days.	D.	M.	S.	M.	S.	Days.	D.	M.	S.	M.	S.	Days.	D.	M.	S.	M.	S.
2800	154	43	54,2			3250,0	156	1	30,4			3750,0	157	11	55,9	1	35,6
2810	154	45	48,8	1	54,6	3262,5	156	3	26,8	1	56,4	3762,5	157	13	31,5		35,1
2820	154	47	42,8		54,0	3275,0	156	5	22,6		55,8	3775,0	157	15	6,6		35,1
2830	154	49	36,2		53,1	3287,5	156	7	17,8		55,2	3787,5	157	16	41,2		31,6
				1	52,9					1	54,6					1	34,2
2840	154	51	29,1		52,3	3300,0	156	9	12,4		54,0	3800,0	157	18	15,4		33,8
2850	154	53	21,4		51,8	3312,5	156	11	6,4		53,3	3812,5	157	19	49,2		33,4
2860	154	55	13,2		51,3	3325	0	12	59,7		52,8	3825,0	157	21	22,6		33,4
2870	154	57	4,5			3337,5	156	14	52,5		52,8	3837,5	157	22	55,6		33,0
				1	50,7					1	52,2					1	32,6
2880	154	58	55,2		50,1	3350,0	156	16	44,7		51,7	3850,0	157	24	28,2		32,1
2890	155	0	45,3		49,7	3362,5	156	18	36,4		51,0	3862,5	157	26	0,3		31,7
2900	155	2	35,0		49,1	3375,0	456	20	27,4		50,5	3875,0	157	27	32,0		31,3
2910	155	4	24,1			3387,5	156	22	17,9			3887,5	157	29	3,3		31,3
				1	48,6					1	50,0					1	31,0
2920	155	6	12,7		48,0	3400,0	156	24	7,9		49,4	3900,0	157	30	34,3		30,5
2930	155	8	0,7		47,6	3412,5	156	25	57,3		48,8	3912,5	157	32	4,8		30,1
2940	155	9	48,3		47,1	3425,0	156	27	46,1		48,2	3925,0	157	33	34,9		29,7
2950	155	11	35,4			3437,5	156	29	34,3			3937,5	157	35	4,6		29,7
				1	46,6					1	47,7					1	29,3
2960	155	13	22 0		46,0	3450,0	156	31	22,0		47,2	3950,0	157	36	33,9		28,9
2970	155	15	8,0		45,6	3462,5	156	33	9,2		46,7	3962,5	157	38	2,8		28,6
2980	155	16	53,6		45,1	3475,0	156	34	55,9		46,1	3975,0	157	39	31,4		28,1
2990	155	18	38,7			3487,5	156	36	42,0			3987,5	157	40	59,5		28,1
				1	44,5					1	45,6					1	27,8
3000,	155	20	23,2	2	10,1	3500,0	156	38	27,6		45,1	4000	157	42	27,3	2	54,5
3012,5	155	22	33,3		9,3	3512,5	156	40	12,7		44,6	4025	157	45	21,8		52,9
3025,0	155	24	42,6		8,6	3525,0	156	41	57,3		44,0	4050	157	48	14,7		51,5
3037,5	155	26	51,2			3537,5	156	43	41,3			4075	157	51	6,2		51,5
				2	7,8					1	43,6					2	50,1
3050,0	155	28	59,0		7,1	3550,0	156	45	24,9		43,0	4100	157	53	56,3		48,6
3062,5	155	31	6,1		6,4	3562,5	156	47	7,9		42,5	4125	157	56	44,9		47,2
3075,0	155	33	12,5		5,7	3575,0	156	48	50,4		42,1	4150	157	59	32,1		45,8
3087,5	155	35	18,2			3587,5	156	50	32,5			4175	158	2	17,9		45,8
				2	5,0					1	41,5					2	44,5
3100,0	155	37	23,2		4,3	3600,0	156	52	14,0		41,1	4200	158	5	2,4		43,1
3112,5	155	39	27,5		3,6	3612,5	156	53	55,1		40,6	4225	158	7	45,5		41,8
3125,0	155	41	31,1		2,9	3625 0	156	55	35,7		40,1	4250	158	10	27,3		40,5
3137,5	155	43	34,0			3637,5	156	57	15,8			4275	158	13	7,8		40,5
				2	2,2					1	39,6					2	39,2
3150,0	155	45	36,2		1,5	3650,0	156	58	55,4		39,2	4300	158	15	47,0		37,9
3162,5	155	47	37,7		0,9	3662,5	157	0	34,6		38,7	4325	158	18	24,9		36,7
3175,0	155	49	38,6		0,3	3675,0	157	2	13,3		38,2	4350	158	21	1,6		35,5
3187,5	155	51	38,9			3687,5	157	3	51,5			4375	158	23	37,1		35,5
				1	59,6					1	37,8					2	34,2
3200,0	155	53	38,5		58,9	3700,0	157	5	29,3		37,3	4400	158	26	11,3		33,0
3212,5	155	55	37,4		58,3	3712,5	157	7	6,6		36,9	4425	158	28	44,3		31,9
3225,0	155	57	35,7		57,7	3725,0	157	8	43,5		36,4	4450	158	31	16,2		30,7
3237,5	155	59	33,4			3737,5	157	10	19,9			4475	158	33	46,9		30,7
				1	57,0					1	36,0					2	29,5

TABLE III. *continued.*

Days.	D. M. S.	M. S.	Days.	D. M. S.	M. S.	Days.	D. M. S.	M. S.
4500	158 36 16,4		5500	160 2 39,3		6500	161 9 49,1	
4525	158 38 44,7	2 28,3	5525	160 4 32,0	1 52,7	6525	161 11 18,8	1 29,7
4550	158 41 12,0	27,3	5550	160 6 24,0	52,0	6550	161 12 48,1	29,3
4575	158 43 38,1	26,1	5575	160 8 15,3	51,3	6575	161 14 16,9	28,8
		2 25,1			1 50,7			1 28,3
4600	158 46 3,2		5600	150 10 6,0		6600	161 15 45,2	
4625	158 48 27,2	24,0	5625	160 11 56,0	50,0	6625	161 17 13,1	27,9
4650	158 50 50,1	22,9	5650	160 13 45,3	49,3	6650	161 18 40,5	27,4
4675	158 53 11,9	21,8	5675	160 15 33,9	48,6	6675	161 20 7,5	27,0
		2 20,8			1 48,0			1 26,5
4700	158 55 32,7		5700	160 17 21,9		6700	161 21 34,0	
4725	158 57 52,5	19,8	5725	160 19 9,3	47,4	6725	161 23 -,1	26,1
4750	159 0 11,3	18,8	5750	160 20 56,0	46,7	6750	161 24 25,7	25,6
4775	159 2 29,1	17,8	5775	160 22 42,0	46,0	6775	161 25 50,9	25,2
		2 16,8			1 45,5			1 24,8
4800	159 4 45,9		5800	160 24 27,5		6800	161 27 15,7	
4825	159 7 1,7	15,8	5825	160 26 12,3	44,8	6825	161 28 40,1	24,4
4850	159 9 16,5	14,8	5850	160 27 56,5	44,2	6850	161 30 4,0	23,9
4875	159 11 30,4	13,9	5875	160 29 40,1	43,6	6875	161 31 27,6	23,6
		2 13,0			1 43,0			1 23,1
4900	159 13 43,4		5900	160 31 23,1		6900	161 32 50,7	
4925	159 15 55,4	12,0	5925	160 33 5,5	42,4	6925	161 34 13,4	22,7
4950	159 18 6,5	11,1	5950	160 34 47,3	41,8	6950	161 35 35,7	22,3
4975	159 20 16,7	10,2	5975	160 36 28,6	41,3	6975	161 36 57,6	21,9
		2 9,3			1 40,6			1 21,4
5000	159 22 26,0		6000	160 38 9,2		7000	161 38 19,0	
5025	159 24¼ 34,4	8,4	6025	160 39 49,3	40,1	7050	161 41 0,8	2 41,8
5050	159 26 42,0	7,6	6050	160 41 28,8	39,5	7100	161 44 41,1	40,3
5075	159 28 48,7	6,7	6075	160 43 7,8	39,0	7150	161 46 19,8	38,7
		2 5,8			1 38,4			2 37,1
5100	159 30 54,5		6100	160 44 46,2		7200	161 48 56,9	
5125	159 32 59,5	5,0	6125	160 46 24,0	37,8	7250	161 51 32,6	35,7
5150	159 35 3,7	4,2	6150	160 48 1,3	37,3	7300	161 54 6,9	34,3
5175	159 37 7,0	3,3	6175	160 49 38,1	36,8	7350	161 56 39,7	32,8
		2 2,5			1 36,2			2 31,4
5200	159 39 9,5		6200	160 51 14,3		7400	161 59 11,1	
5225	159 41 11,2	1,7	6225	160 52 50,0	35,7	7450	162 1 41,1	30,0
5250	159 43 12,1	0,9	6250	160 54 25,2	35,2	7500	162 4 9,8	28,7
5275	159 45 12,2	0,1	6275	160 55 59,8	34,6	7550	162 6 37,1	27,3
		1 59,4			1 34,2			2 25,9
5300	159 47 11,6		6300	160 57 34,0		7600	162 9 3,0	
5325	159 49 10,2	58,6	6325	160 59 7,6	33,6	7650	162 11 27,7	24,7
5350	159 51 8,0	57,8	6350	161 0 40,7	33,1	7700	162 13 51,1	23,4
5375	159 53 5,0	57,0	6375	161 2 13,3	32,6	7750	162 16 13,2	22,1
		1 56,3			1 32,2			2 20,9
5400	159 55 1,3		6400	161 3 45,5		7800	162 18 34,1	
5425	159 56 56,9	55,6	6425	161 5 17,1	31,6	7850	162 20 53,7	19,6
5450	159 58 51,8	54,9	6450	161 6 48,3	31,2	7900	162 23 12,2	18,5
5475	160 0 45,9	54,1	6475	161 8 18,9	30,6	7950	162 25 29,4	17,2
		1 53,4			1 30,2			2 16,1

TABLE III. *continued.*

Days.	D. M. S.	M. S.	Days.	D. M. S.	M. S.	Days.	D. M. S.	M. S.
8000	162 27 45,5		10000	163 45 13,4		12000	164 44 3,8	
8050	162 30 0,4	2 14,9	10050	163 46 53,0	1 39,6	12100	164 46 39,1	2 35,3
8100	162 32 14,2	13,8	10100	163 48 32,0	39,0	12200	164 49 12,7	33,6
8150	162 34 26,9	12,7	10150	163 50 10,4	38,4	12300	164 51 44,6	31,9
8200	162 36 38,5	2 11,6	10200	163 51 48,1	1 37,7	12400	164 54 14,8	2 30,2
8250	162 38 49,0	10,5	10250	163 53 25,1	37,0	12500	164 56 43,4	28,6
8300	162 40 58,4	9,4	10300	163 55 1,5	36,4	12600	164 59 10,8	26,9
8350	162 43 6,7	8,3	10350	163 56 37,2	35,7	12700	165 1 35,7	25,4
8400	162 45 14,0	2 7,3	10400	163 58 12,4	1 35,2	12800	165 3 59,6	2 23,9
8450	162 47 20,3	6,3	10450	163 59 46,9	34,5	12900	165 6 21,9	22,3
8500	162 49 25,6	5,3	10500	164 1 20,8	33,9	13000	165 8 42,8	20,9
8550	162 51 29,8	4,2	10550	164 2 54,1	33,8	13100	165 11 2,2	19,4
8600	162 53 33,1	2 3,3	10600	164 4 26,8	1 32,7	13200	165 13 20,2	2 18,0
8650	162 55 35,4	2,3	10650	164 5 58,9	32,1	13300	165 15 36,7	16,5
8700	162 57 36,8	1,4	10700	164 7 30,4	31,5	13400	165 17 51,9	15,2
8750	162 59 37,2	0,1	10750	164 . 9 1,4	31,0	13500	165 20 5,7	13,8
8800	163 1 36,6	1 59,4	10800	164 10 31,7	1 30,3	13600	165 22 18,2	2 12,5
8850	163 3 35,2	58,6	10850	164 12 1,5	29,8	13700	165 24 29,4	11,2
8900	163 5 32,8	57,6	10900	164 13 30,8	29,3	13800	165 26 39,2	9,8
8950	163 7 29,6	56,8	10950	164 14 59,5	28,7	13900	165 28 47,8	8,6
9000	163 9 25,4	1 55,8	11000	164 16 27,6	1 28,1	14000	165 30 55,2	2 7,4
9050	163 11 20,4	55,-	11050	164 17 55,2	27,6	14100	165 33 1,3	6,1
9100	163 13 14,5	54,1	11100	164 19 22,3	27,1	14200	165 35 6,3	5,0
9150	163 15 7,8	53,3	11150	164 20 48,8	26,5	14300	165 37 10,0	3,7
9200	163 17 0,3	1 52,5	11200	164 22 14,8	1 26,0	14400	165 39 12,6	2 2,6
9250	163 18 51,9	51,6	11250	164 23 40,3	25,5	14500	165 41 14,0	1,4
9300	163 20 42,7	50,8	11300	164 25 5,3	25,0	14600	165 43 14,3	0,3
9350	163 22 32,7	50,0	11350	164 26 29,7	24,4	14700	165 45 13,5	1 59,2
9400	163 24 21,8	1 49,1	11400	164 27 53,7	1 24,0	14800	165 47 11,6	1 58,1
9450	163 26 10,2	48,4	11450	164 29 17,2	23,5	14900	165 49 8,6	57,0
9500	163 27 57,9	47,7	11500	164 30 40,1	22,9	15000	165 51 4,6	56,0
9550	163 29 44,7	46,8	11550	164 32 2,6	22,5	15100	165 52 59,5	54,9
9600	163 31 30,8	1 46,1	11600	164 33 24,6	1 22,0	15200	165 54 53,4	1 53,9
9650	163 33 16,2	45,4	11650	164 34 46,1	21,5	15300	165 56 46,3	52,9
9700	163 35 0,8	44,6	11700	164 36 7,2	21,1	15400	165 58 38,2	51,9
9750	163 36 44,6	43,8	11750	164 37 27,8	20,6	15500	166 0 29,2	51,0
9800	163 38 27,8	1 43,2	11800	164 38 47,9	1 20,1	15600	166 2 19,1	1 49,9
9850	163 40 10,2	42,4	11850	164 40 7,5	19,6	15700	166 4 8,1	49,0
9900	163 41 52,0	41,8	11900	164 41 26,7	19,-	15800	166 5 56,2	48,1
9950	163 43 33,0	41,0	11950	164 42 45,5	18,8	15900	166 7 43,4	47,2
		1 40,4			1 18,3			1 16,2

TABLE III. continued.

Days.	D.	M.	S.	M.	S.	Days.	D.	M.	S.	M.	S.	Days.	D.	M.	S.	M.	S.
16000	166	9	29,6			20000	167	10	2,4			24000	167	56	7,2		
16100	166	11	15,0	1	45,4	20100	167	11	20,4	1	18,0	24200	167	58	9,0	2	1,8
16200	166	12	59,5		44,5	20200	167	12	37,9		17,5	24400	168	0	9,4		0,4
16300	166	14	43,1		43,6	20300	167	13	54,9		17,0	24600	168	2	8,4	1	59,0
16400	166	16	25,8	1	42,7	20400	167	15	11,3	1	16,4	24800	168	4	6,2	1	57,8
16500	166	18	7,7		41,9	20500	167	16	27,3		16,0	25000	168	6	2,7		56,5
16600	166	19	48,8		41,1	20600	167	17	42,8		15,5	25200	168	7	57,9		55,2
16700	166	21	29,0		40,2	20700	167	18	57,7		14,9	25400	168	9	51,9		54,0
16800	166	23	8,5	1	39,5	20800	167	20	12,2	1	14,5	25600	168	11	44,7	1	52,8
16900	166	24	47,1		38,6	20900	167	21	26,2		14,0	25800	168	13	36,4		51,7
17000	166	26	25,0		37,9	21000	167	22	39,7		13,5	26000	168	15	26,9		50,5
17100	166	28	2,1		37,1	21100	167	23	52,7		13,0	26200	168	17	16,2		49,3
17200	166	29	38,4	1	36,3	21200	167	25	5,3	1	12,6	26400	168	19	4,4	1	48,2
17300	166	31	14,0		35,6	21300	167	26	17,4		12,1	26600	168	20	51,5		47,1
17400	166	32	48,8		34,8	21400	167	27	29,1		11,7	26800	168	22	37,5		46,0
17500	166	34	22,9		34,1	21500	167	28	40,3		11,2	27000	168	24	22,5		45,0
17600	166	35	56,3	1	33,4	21600	167	29	51,1	1	10,8	27200	168	26	6,4	1	43,9
17700	166	37	29,0		32,7	21700	167	31	1,4		10,3	27400	168	27	49,3		42,9
17800	166	39	0,9		31,9	21800	167	32	11,3		9,9	27600	168	29	31,2		41,9
17900	166	40	32,2		31,0	21900	167	33	20,7		9,4	27800	168	31	12,1		40,9
18000	166	42	2,8	1	30,6	22000	167	34	29,7	1	9,0	28000	168	32	52,1	1	40,0
18100	166	43	32,6		29,8	22100	167	35	38,3		8,6	28200	168	34	31,1		39,0
18200	166	45	1,8		29,2	22200	167	36	46,5		8,2	28400	168	36	9,1		38,0
18300	166	46	30,4		28,6	22300	167	37	54,3		7,8	28600	168	37	46,2		37,1
18400	166	47	58,3	1	27,9	22400	167	39	1,7	1	7,4	28800	168	39	22,4	1	36,2
18500	166	49	25,6		27,3	22500	167	40	8,6		6,9	29000	168	40	57,7		35,3
18600	166	50	52,2		26,6	22600	167	41	15,2		6,6	29200	168	42	32,1		34,4
18700	166	52	18,2		26,0	22700	167	42	21,4		6,2	29400	168	44	5,6		33,5
18800	166	53	43,6	1	25,4	22800	167	43	27,1	1	5,7	29600	168	45	38,3	1	32,7
18900	166	55	8,3		24,7	22900	167	44	32,5		5,4	29800	168	47	10,2		31,9
19000	166	56	32,5		24,2	23000	167	45	37,5		5,0	30000	168	48	41,3		31,1
19100	166	57	56,0		23,5	23100	167	46	42,1		4,6	30200	168	50	11,5		30,2
19200	166	59	19,0	1	23,0	23200	167	47	46,4	1	4,3	30400	168	51	40,9	1	29,4
19300	167	0	41,4		22,4	23300	167	48	50,3		3,9	30600	168	53	9,5		28,6
19400	167	2	3,2		21,8	23400	167	49	53,8		3,5	30800	168	54	37,3		27,8
19500	167	3	24,5		21,3	23500	167	50	56,9		3,1	31000	168	56	4,4		27,1
19600	167	4	45,2	1	20,7	23600	167	51	59,7	1	2,8	31200	168	57	30,8	1	26,4
19700	167	6	5,3		20,1	23700	167	53	2,1		2,4	31400	168	58	56,4		25,6
19800	167	7	24,9		19,6	23800	167	54	4,2		2,1	31600	169	0	21,3		24,9
19900	167	8	43,9		19,0	23900	167	55	5,9		1,7	31800	169	1	45,4		24,1
				1	18,5					1	1,3					1	23,5

TABLE III. *continued.*

Days.	D. M. S.	M. S.	Days.	D. M. S.	M. S.	Days.	D. M. S.	M. S.
32000	169 3 8,9		40000	169 50 44,3		52000	170 42 13,2	
32200	169 4 31,6	1 22,7	40250	169 52 0,9	1 16,6	52500	170 44 0,8	1 47,6
32400	169 5 53,7	22,1	40500	169 53 16,9	16,0	53000	170 45 46,9	46,1
32600	169 7 15,0	21,3	40750	169 54 32,3	15,4	53500	170 47 31,7	44,8
		1 20,7			1 14,8			1 43,5
32800	169 8 35,7	20,1	41000	169 55 47,1	14,1	54000	170 49 15,2	42,2
33000	169 9 55,8	19,4	41250	169 57 1,2	13,6	54500	170 50 57,4	41,0
33200	169 11 15,2	18,7	41500	169 58 14,8	12,9	55000	170 52 38,4	39,7
33400	169 12 33,9		41750	169 59 27,7		55500	170 54 18,1	
		1 18,2			1 12,4			1 38,6
33600	169 13 52,1	17,5	42000	170 0 40,1	11,8	56000	170 55 56,7	37,4
33800	169 15 9,6	16,9	42250	170 1 51,9	11,2	56500	170 57 34,1	36,2
34000	169 16 26,5	16,2	42500	170 3 3,1	10,7	57000	170 59 10,3	35,1
34200	169 17 42,7		42750	170 4 13,8		57500	171 0 45,4	
		1 15,7			1 10,1			1 34,0
34400	169 18 58,4	15,1	43000	170 5 23,9	9,6	58000	171 2 19,4	32,9
34600	169 20 13,5	14,5	43250	170 6 33,5	9,0	58500	171 3 52,3	31,9
34800	169 21 28,0	14,0	43500	170 7 42,5	8,5	59000	171 5 24,2	30,8
35000	169 22 42,0		43750	170 8 51,0		59500	171 6 55,0	
		1 13,4			1 7,9			1 29,8
35200	169 23 55,4	12,8	44000	170 9 58,9	7,4	60000	171 8 24,8	28,8
35400	169 25 8,2	12,3	44250	170 11 6,3	6,9	60500	171 9 53,6	27,8
35600	169 26 20,5	11,7	44500	170 12 13,2	6,5	61000	171 11 21,4	26,9
35800	169 27 32,2		44750	170 13 19,7		61500	171 12 48,3	
		1 11,1			1 5,9			1 25,9
36000	169 28 43,3	10,7	45000	170 14 25,6	5,5	62000	171 14 14,2	25,0
36200	169 29 54,0	10,1	45250	170 15 31,1	4,9	62500	171 15 39,2	24,1
36400	169 31 4,1	9,7	45500	170 16 36,0	4,5	63000	171 17 3,0	23,2
36600	169 32 13,8		45750	170 17 40,5		63500	171 18 26,5	
		1 9,1			1 4,0			1 22,3
36800	169 33 22,9	8,6	46000	170 18 44,5	3,5	64000	171 19 48,8	21,5
37000	169 34 31,5	8,1	46250	170 19 48,0	3,0	64500	171 21 10,3	20,6
37200	169 35 39,6	7,6	46500	170 20 51,0	2,6	65000	171 22 30,9	19,8
37400	169 36 47,2		46750	170 21 53,6		65500	171 23 50,7	
		1 7,1			1 2,2			1 19,0
37600	169 37 54,3	6,7	47000	170 22 55,8	1,7	66000	171 25 9,7	18,2
37800	169 39 1,0	6,2	47250	170 23 57,5	1,3	66500	171 26 27,9	17,4
38000	169 40 7,2	5,7	47500	170 24 58,8	0,9	67000	171 27 45,3	16,6
38200	169 41 12,9		47750	160 25 59,7		67500	171 29 1,9	
		1 5,3			1 0,4			1 15,9
38400	169 42 18,2	4,8	48000	170 27 0,1	59,6	68000	171 30 17,8	15,1
38600	169 43 23,0	4,3	48500	170 28 59,7	58,0	68500	171 31 32,9	14,4
38800	169 44 27,3	3,9	49000	170 30 57,7	56,4	69000	171 32 47,3	13,7
39000	169 45 31,2		49500	170 32 54,1		69500	161 34 1,0	
		1 3,5			1 54,8			1 13,0
39200	169 46 34,7	3,1	50000	170 34 48,9	53,3	70000	171 35 14,0	12,3
39400	169 47 37,8	2,6	50500	170 36 42,2	51,8	70500	171 36 26,3	11,6
39600	169 48 40,4	2,2	51000	170 38 34,0	50,3	71000	171 37 37,9	10,9
39800	169 49 42,6		51500	170 40 24,3		71500	171 38 48,8	
		1 1,7			1 48,9			1 10,3

TABLE III. *continued*.

Days.	D.	M.	S.	M.	S.	Days.	D.	M.	S.	M.	S.	Days.	D.	M.	S.	M.	S.
72000	171	39	59,1			92000	172	19	27,5	0	50,2	124000	173	3	17,8	1	7,3
72500	171	41	8,7	1	9,6	92500	172	20	17,7		49,8	125000	173	4	25,1		6,4
73000	171	42	17,6		8,9	93000	172	21	7,5		49,4	126000	173	5	31,5		5,8
73500	171	43	25,9		8,3	93500	172	21	56,9			127000	173	6	37,3		
				1	7,8					0	49,1					1	5,1
74000	171	44	33,7		7,1	94000	172	22	46,0		48,7	128000	173	7	42,4		4,3
74500	171	45	40,8		6,5	94500	172	23	34,7		48,5	129000	173	8	46,7		3,8
75000	171	46	47,3		5,9	95000	172	24	23,2		48,0	130000	173	9	50,5		3,1
75500	171	47	53,2			95500	172	25	11,2			131000	173	10	53,6		
				1	5,3					0	47,7					1	2,1
76000	171	48	58,5		4,8	96000	172	25	58,9		47,4	132000	173	11	56,0		1,9
76500	171	50	3,3		4,2	96500	172	26	46,3		47,1	133000	173	12	57,9		1,1
77000	171	51	7,5		3,6	97000	172	27	33,4		46,8	134000	173	13	59,0		0,5
77500	171	52	11,1			97500	172	28	20,2			135000	173	14	59,5		
				1	3,1					0	46,4					1	0,1
78000	171	53	14,2		2,5	98000	172	29	6,6		46,1	136000	173	15	59,6	0	59,4
78500	171	54	16,7		2,1	98500	172	29	52,7		45,8	137000	173	16	59,0		58,8
79000	171	55	18,8		1,5	99000	172	30	38,5		45,5	138000	173	17	57,8		58,2
79500	171	56	20,3			99500	172	31	24,0			139000	173	18	56,0		
				1	1,0					0	45,2					0	57,7
80000	171	57	21,3		0,4	100000	172	32	9,2	1	29,5	140000	173	19	53,7		57,1
80500	171	58	21,7		0,0	101000	172	33	38,7		29,1	141000	173	20	50,8		56,6
81000	171	59	21,7	0	59,5	102000	172	35	7,8		26,3	142000	173	21	47,4		56,1
81500	172	0	21,2			103000	172	36	34,1			143000	173	22	43,5		
				0	59,0					1	26,0					0	55,5
82000	172	1	20,2		58,5	104000	172	38	0,1		25,0	144000	173	23	39,0		55,1
82500	172	2	18,7		58,0	105000	172	39	25,1		23,8	145000	173	24	34,1		54,5
83000	172	3	16,7		57,6	106000	172	40	48,9		22,8	146000	173	25	28,6		54,0
83500	172	4	14,3			107000	172	42	11,7			147000	173	26	22,6		
				0	57,1					1	21,8					0	53,6
84000	172	5	11,4		56,7	108000	172	43	33,5		20,7	148000	173	27	16,2		53,1
84500	172	6	8,1		56,2	109000	172	44	54,2		19,8	149000	173	28	9,3		52,5
85000	172	7	4,3		55,8	110000	172	46	14,0		18,8	150000	173	29	1,8		52,2
85500	172	8	0,1			111000	172	47	32,8			151000	173	29	54,0		
				0	55,3					1	17,8					0	51,8
86000	172	8	55,4		54,9	112000	172	48	50,6		17,0	152000	173	30	45,8		51,0
86500	172	9	50,3		54,5	113000	172	50	7,6		16,0	153000	173	31	36,8		50,8
87000	172	10	44,8		54,1	114000	172	51	23,6		15,1	154000	173	32	27,6		50,3
87500	172	11	38,9			115000	172	52	38,7			155000	173	33	17,9		
				0	53,6					1	14,3					0	49,9
88000	172	12	32,5		53,2	116000	172	53	53,0		13,1	156000	173	34	7,8		49,5
88500	172	13	25,7		52,9	117000	172	55	6,4		12,6	157000	173	34	57,3		49,0
89000	172	14	18,6		52,1	118000	172	56	19,0		11,8	158000	173	35	46,3		48,6
89500	172	15	11,0			119000	172	57	30,8			159000	173	36	34,0		
				0	52,1					1	10,9					0	48,0
90000	172	16	3,1		51,6	120000	172	58	41,7		10,2	160000	173	37	23,0		47,9
90500	172	16	54,7		51,3	121000	172	59	51,9		9,4	161000	173	38	11,1		47,6
91000	172	17	46,0		50,9	122000	173	1	1,3		8,6	162000	173	38	58,7		47,2
91500	172	18	36,9			123000	173	2	9,9			163000	173	39	45,9		
				0	49,6					1	7,9					0	46,7

TABLE III. *continued.*

Days.	D.	M.	S.	M.	S.
164000	173	40	32,6		
165000	173	41	19,0	o	46,4
166000	173	42	5,0		46,0
167000	173	42	50.6		45,6
				o	45,3
168000	173	43	35,9		
169000	173	44	20,8		44,9
170000	173	45	5,4		44,6
171000	173	45	49,6		44,2
				o	43,9
172000	173	46	33,5		
173000	173	47	17,0		43,5
174000	173	48	0,1		43,1
175000	173	48	42,9		42,8
				o	42,4

Days.	D.	M.	S.	M.	S.
176000	173	49	25,3		
177000	173	50	7,3	o	42,0
178000	173	50	49,0		41,7
179000	173	51	30,3		41,3
				o	41,0
180000	173	52	11,3		
181000	173	52	52,0		40,7
182000	173	53	32,4		40,4
183000	173	54	12,5		40,1
				o	39,8
184000	173	54	52,3		
185000	173	55	31,9		39,6
186000	173	56	11,2		39,3
187000	173	56	50,3		39,1
				o	38,9

Days.	D.	M.	S.	M.	S.
188000	173	57	29,2		
189000	173	58	7,7	o	38,5
190000	173	58	46,0		38,3
191000	173	59	24,1		38,1
				o	37,8
192000	174	0	1,9		
193000	174	0	39,4		37,5
194000	174	1	16,6		37,2
195000	174	1	53,6		37,0
				o	36,7
196000	174	2	30,3		
197000	174	3	6,8		36,5
198000	174	3	43,1		36,3
199000	174	4	19,1		36,0
				o	35,7
200000	174	4	54,8		

3 N

TABLE IV. Barker's *general Table of the Parabola.*

Angle from Perihelion.	Mean Mot.	Diff.	Log. Dift.	Diff.		Angle.	Mean Mot.	Diff.	Log. Dift.	Diff.
°		−545		0		°		−546		20
o 5	0.0545	546	0.000000	1		35	2,3468	547	0.000425	20
10	0.1091	545	0.000001	1		40	2,4015	546	0.000445	20
15	0.1636	546	0.000002	2		45	2,4561	547	0.000465	21
20	0.2182	545	0.000004	2		50	2,5108	546	0.000486	21
25	0 2727	545	0.000006	2		55	2,5654	547	0.000507	22
30	0 3272		0.000008			4 0	2,6201		0,000529	
		−546		3				−547		23
35	0.3818	545	0.000011	4		5	2,6748	547	0.000552	22
40	0.4363	546	0.000015	4		10	2,7295	547	0.000574	23
45	0.4909	545	00000.19	4		15	2,7842	547	0.000597	24
50	0.5454	546	0.000023	5		20	2,8389	547	0.000621	24
55	0 5000	545	0.000028	5		25	2,8936	547	0.000645	25
1 0	0 6545		0.000033			30	2,9483		0.000670	
		−546		6				−547		25
5	0.7091	545	0.000039	6		4 35	3,0030	547	0.000695	25
10	0.7636	546	0.000045	7		40	3,0577	547	0.000720	26
15	0.8182	545	0.000052	7		45	3,1124	548	0.000746	27
20	0.8727	546	0.000059	7		50	3,1672	548	0.000773	27
25	0.9273	546	0.000066	8		55	3,2219	547	0.000800	27
30	0.9819		0.000074			5 0	3,2766		0.000827	
		−545		9				−548		28
35	1.0364	546	0.000083	9		5	3,3314	548	0.000855	28
40	1.0910	546	0.000092	9		10	3,3862	547	0.000883	29
45	1.1456	545	0.000101	10		15	3,4409	548	0.000912	29
50	1.2001	546	0.000111	10		20	3,4957	548	0.000941	29
55	1.2547	546	0.000121	11		25	3,5505	548	0.000971	30
2 0	1.3093		0.000132			30	3,6053	548	0.001001	
		−545		12				−548		30
5	1.3638	546	0.000144	11		35	3,6601	548	0.001031	30
10	1.4184	546	0.000155	12		40	3,7149	548	0.001062	31
15	1.4730	546	0.000167	13		45	3,7697	548	0.001094	32
20	1.5276	546	0.000180	13		50	3,8245	548	0.001126	32
25	1.5822	546	0.000193	14		55	3,8793	548	0.001158	32
30	1.6368		0.000207			6 0	3,9342	549	0.001191	
		−546		14				−548		33
35	1.6914	546	0.000221	14		5	3,9890	548	0.001224	33
40	1.7460	546	0.000235	15		10	4,0439	549	0.001258	34
45	1.8006	546	0.000250	16		15	4,0988	549	0.001293	35
50	1.8552	546	0.000266	15		20	4,1536	548	0.001327	34
55	1.9098	546	0.000281	17		25	4,2085	549	0.001362	35
3 0	1.9644		0.000298			30	4,2634	549	0.001398	36
		−546		16				−549		36
5	2.0190	546	0.000314	18		35	4,3183	549	0.001434	37
10	2.0736	547	0.000332	17		40	4,3732	549	0.001471	37
15	2.1283	546	0.000349	19		45	4,4281	549	0.001508	37
20	2.1829	546	0.000368	18		50	4,4830	549	0.001545	38
25	2.2375	547	0.000386	19		55	4,5380	550	0.001583	39
30	2.2922	546	0.000405	20		7 0	4,5929	550	0.001622	39

TABLE IV. *continued.*

Angle.		Mean Mot.	Diff.	Log. Diff.	Diff.	Angle.		Mean Mot.	Diff.	Log. Diff.	Diff.
°	′		—550		—39	°			—555		—58
7	5	4.6479		0.001661		10	35	6.9664		0.003710	
	10	4.7028	549	0.001700	39		40	7.0219	555	0.003769	59
	15	4.7578	550	0.001740	40		45	7.0774	555	0.003828	59
	20	4.8128	550	0.001780	40		50	7.1329	555	0.003887	59
	25	4.8678	550	0.001821	41		55	7.1885	556	0.003947	60
	30	4.9228	550	0.001862	41	11	0	7.2440	555	0.004008	61
			—550		—11				—556		—61
	35	4.9778	550	0.001903	42		5	7.2996	555	0.004069	62
	40	5.0328	551	0.001945	43		10	7.3551	555	0.004131	62
	45	5.0879	550	0.001988	43		15	7.4107	556	0.004193	62
	50	5.1429	551	0.002031	43		20	7.4663	556	0.004255	63
	55	5.1980	551	0.002074	44		25	7.5220	557	0.004318	63
8	0	5.2531	551	0.002118			30	7.5776	556	0.004381	
			550		—15				—557		—64
	5	5.3081	551	0.002163	45		35	7.6333	557	0.004445	64
	10	5.3632	551	0.002208	45		40	7.6890	557	0.004509	65
	15	5.4183	551	0.002253	46		45	7.7447	557	0.004574	65
	20	5.4734	552	0.002299	46		50	7.8004	557	0.004639	66
	25	5.5286	551	0.002345	47		55	7.8561	557	0.004705	66
	30	5.5837		0.002392		12	0	7.9118		0.004771	
			—552		—17				—558		—67
	35	5.6389	551	0.002439	48		5	7.9676	558	0.004838	67
	40	5.6940	552	0.002487	48		10	8.0234	558	0.004905	68
	45	5.7492	552	0.002535	48		15	8.0792	558	0.004973	68
	50	5.8044	552	0.002583	49		20	8.1350	558	0.005041	68
	55	5.8596	552	0.002632	50		25	8.1908	559	0.005109	69
9	0	5.9148	552	0.002682			30	8.2467		0.005178	
			—552		—50				—558		—69
	5	5.9700	553	0.002732	50		35	8.3025	559	0.005247	70
	10	6.0253	552	0.002782	51		40	8.3584	559	0.005317	71
	15	6.0805	553	0.002833	51		45	8.4143	559	0.005388	70
	20	6.1358	553	0.002884	52		50	8.4702	560	0.005458	72
	25	6.1911	553	0.002936	52		55	8.5262	560	0.005530	72
	30	6.2464		0.002988		13	0	8.5822	560	0.005602	
			—553		—53				—559		—72
	35	6.3017	553	0.003041	53		5	8.6381	560	0.005674	72
	40	6.3570	553	0.003094	53		10	8.6941	560	0.005746	72
	45	6.4123	554	0.003148	54		15	8.7501	561	0.005819	73
	50	6.4677	553	0.003202	54		20	8.8062	561	0.005893	74
	55	6.5230	554	0.003257	55		25	8.8622	560	0.005967	74
10	0	6.5784		0.003312			30	8.9183	561	0.006042	75
			—554		—55				—561		—75
	5	6.6338	554	0.003367	56	13	35	8.9744	561	0.006117	75
	10	6.6892	554	0.003423	56		40	9.0305	561	0.006192	76
	15	6.7446	554	0.003479	57		45	9.0866	562	0.006268	76
	20	6.8000	554	0.003536	58		50	9.1428	561	0.006344	77
	25	6.8555	555	0.003594	58		55	9.1989	562	0.006421	77
	30	6.9109	554	0.003652		14	0	9.2551	562	0.006498	
			—555		—58				—562		—78

TABLE IV. *continued.*

Angle.		Mean Mot.	Diff.	Log. Diff.	Diff.
°			−562		−78
14	5	9.3113		0.006576	
	10	9.3676	563	0.006655	79
	15	9.4238	562	0.006733	78
	20	9.4801	563	0.006813	80
	25	9.5364	563	0.006892	79
	30	9.5927	563	0.006972	80
			−563		−81
	35	9.6490	564	0.007053	81
	40	9.7054	563	0.007134	81
	45	9.7617	564	0.007216	82
	50	9.8181	564	0.007298	82
	55	9.8745	565	0.007380	82
15	0	9.9310		0.007463	83
			−564		−83
	5	9.9874	565	0.007546	84
	10	10.0439	565	0.007630	84
	15	10.1004	565	0.007714	85
	20	10.1569	566	0.007799	85
	25	10.2135	566	0.007884	86
	30	10.2701		0.007970	
			−566		−86
	35	10.3267	566	0.008056	87
	40	10.3833	566	0.008143	87
	45	10.4399	567	0.008230	88
	50	10.4966	567	0.008318	83
	55	10.5533	567	0.008406	88
16	0	10.6100		0.008494	
			−567		−89
	5	10.6667	567	0.008583	90
	10	10.7234	568	0.008673	90
	15	10.7802	568	0.008763	90
	20	10.8370	568	0.008853	91
	25	10.8938	569	0.008944	92
	30	10.9507		0.009036	
			−569		−91
	35	11.0076	569	0.009127	93
	40	11.0645	569	0.009220	92
	45	11.1214	569	0.009312	94
	50	11.1783	570	0.009406	93
	55	11.2353	570	0.009499	94
17	0	11.2923		0.009593	
			−570		−95
	5	11.3493	570	0.009688	95
	10	11.4063	571	0.009783	95
	15	11.4634	571	0.009879	96
	20	11.5205	571	0.009975	96
	25	11.5776	572	0.010071	96
	30	11.6348		0.010168	97
			−571		−98

Angle.		Mean Mot.	Diff.	Log. Diff.	Diff.
°	′		−571		−98
17	35	11.6919		0.010266	
	40	11.7491	572	0.010364	98
	45	11.8063	572	0.010462	98
	50	11.8636	573	0.010561	99
	55	11.9209	573	0.010660	99
18	0	11.9782	573	0.010760	100
			−573		−100
	5	12.0355		0.010800	
	10	12.0928	573	0.010961	101
	15	12.1502	574	0.011062	101
	20	12.2076	574	0.011164	102
	25	12.2650	574	0.011266	102
	30	12.3225	575	0.011369	103
			−575		−103
	35	12.3800		0.011472	
	40	12.4375	575	0.011576	104
	45	12.4951	576	0.011680	104
	50	12.5526	575	0.011784	104
	55	12.6102	576	0.011889	105
19	0	12.6679	577	0.011995	106
			−576		−106
	5	12.7255		0.012101	
	10	12.7832	577	0.012207	106
	15	12.8409	577	0.012314	107
	20	12.8986	577	0.012421	107
	25	12.9564	578	0.012529	108
	30	13.0142	578	0.012637	108
			−578		−109
	35	13.0720		0.012746	
	40	13.1299	579	0.012855	109
	45	13.1878	579	0.012965	110
	50	13.2457	579	0.013075	110
	55	13.3036	579	0.013186	111
20	0	13.3616	580	0.013297	111
			−580		−112
	5	13.4196		0.013409	
	10	13.4776	580	0.013521	112
	15	13.5357	581	0.013633	112
	20	13.5938	581	0.013746	113
	25	13.6519	581	0.013860	114
	30	13.7100	581	0.013974	114
			−582		−114
	35	13.7682		0.014088	
	40	13.8264	582	0.014203	115
	45	13.8847	583	0.014319	116
	50	13.9430	583	0.014435	116
	55	14.0013	583	0.014551	116
21	0	14.0596	583	0.014668	117
			−584		−117

TABLE IV. *continued.*

Angle.		Mean Mot.	Diff.	Log. Dift.	Diff.	Angle.		Mean Mot.	Diff.	Log. Dift.	Diff.
°						°	′				
			−584		−117				−598		−158
21	5	14.1180		0.014785		24	35	16.5998		0.020143	
	10	14.1764	584	0.014903	118		40	16.6597	599	0.020281	138
	15	14.2348	584	0.015021	118		45	16.7196	599	0.020419	158
	20	14.2933	585	0.015140	119		50	16.7795	599	0.020558	139
	25	14.3518	585	0.015259	119		55	16.8395	600.	0.020697	139
	30	14.4103	585	0.015379	120	25	0	16.8995	600	0.020837	140
			−585		−120				−601		−140
	35	14.4688	586	0.015499	121		5	16.9596	600	0.020977	140
	40	14.5274	587	0.015620	121		10	17.0196	602	0.021118	141
	45	14.5861	586	0.015741	121		15	17.0798	602	0.021259	141
	50	14.6447	587	0.015862	122		20	17.1400	602	0.021401	142
	55	14.7034	587	0.015984	123		25	17.2002	602	0.021543	142
22	0	14.7621		0.016107			30	17.2604	602	0.021686	143 ·
			−588		−123				−603		−143
	5	14.8209	588	0.016230	123		35	17.3207	603	0.021829	143
	10	14.8797	588	0.016353	123		40	17.3810	604	0.021973	144
	15	14.9385	588	0.016477	124		45	17.4414	604	0.022117	144
	20	14.9974	589	0.016602	125		50	17.5018	605	0.022261	144
	25	15.0563	589	0.016727	125		55	17.5623	605	0.022407	146
	30	15.1152	589	0.016852	125	26	0	17.6228		0.022552	145
			−589		−126				−605		−146
	35	15.1741	590	0.016978	126		5	17.6833	606	0.022698	146
	40	15.2331	591	0.017104	127		10	17.7439	606	0.022845	147
	45	15.2922	590	0.017231	128		15	17.8045	606	0.022992	147
	50	15.3512	591	0.017359	127		20	17.8651	607	0.023139	147
	55	15.4103	592	0.017486	129		25	17.9258	607	0.023287	148
23	0	15.4695		0.017615			30	17.986ſ		0.023436	149
			−591		−128				−608		−149
	5	15.5286	592	0.017743	130		35	18.0473	608	0.023585	149
	10	15.5878	593	0.017873	129		40	18.1081	609	0.023734	149
	15	15.6471	592	0.018002	130		45	18.1690	609	0.023884	150
	20	15.7063	594	0.018132	131		50	18.2299	610	0.024035	151
	25	15.7657	593	0.018263	131		55	18.2909	610	0.024186	151
	30	15.8250		0.018394		27	0	18.3519		0.024337	151
			−594		−132				−610		−152
	35	15.8844	594	0.018526	132		5	18.4129	611	0.024489	152
	40	15.9438	594	0.018658	132		10	18.4740	611	0.024641	152
	45	16.0032	595	0.018791	133		15	18.5351	611	0.024794	153
	50	16.0627	596	0.018924	133		20	18.5962	612	0.024947	153
	55	16.1223	595	0.019057	133		25	18.6574	613	0.025101	154
24	0	16.1818		0.019191	134		30	18.7187		0.025256	155
			−596		−135				−13		−154
	5	16.2414	597	0.019326	135		35	18.7800	613	0.025410	154
	10	16.3011	596	0.019461	135		40	18.8413	614	0.025566	156
	15	16.3607	597	0.019596	136		45	18.9027	614	0.025722	156
	20	16.4204	598	0.019732	137		50	18.9641	615	0.025878	156
	25	16.4802	598	0.019869	136		55	19.0256	615	0.026035	157
	30	16.5400		0.020005		28	0	19.0871		0.026192	157
			−598		−138				−615		−158

3 O

TABLE IV. continued.

Angle.	Mean Mot.	Diff.	Log. Dist.	Diff.	Angle.	Mean Mot.	Diff.	Log. Dist.	Diff.
		−615		−158			−636		−178
28 5	19.1486	616	0.026350	158	31 35	21.7766	636	0.033417	179
10	19.2102	617	0.026508	158	40	21.8402	637	0.033596	180
15	19.2719	617	0.026666	160	45	21.9039	638	0.033776	179
20	19.3336	617	0.026826	159	50	21.9677	638	0.033955	181
25	19.3953	618	0.026985	160	55	22.0315	638	0.034136	181
30	19.4571		0.027145		32 0	22.0953		0.034317	
		−618		−161			−639		−181
35	19.5189	619	0.027306	161	5	22.1592	640	0.034498	182
40	19.5808	619	0.027467	162	10	22.2232	640	0.034680	182
45	19.6427	620	0.027629	162	15	22.2872	641	0.034862	183
50	19.7047	620	0.027791	163	20	22.3513	641	0.035045	184
55	19.7667	621	0.027954	163	25	22.4154	642	0.035229	183
29 0	19.8288		0.028117		30	22.4796		0.035412	
		−621		−163			−642		−185
5	19.8909	621	0.028280	164	35	22.5438	643	0.035597	185
10	19.9530	622	0.028444	165	40	22.6081	643	0.035782	185
15	20.0152	622	0.028609	165	45	22.6724	644	0.035967	186
20	20.0774	623	0.028774	166	50	22.7368	645	0.036153	186
25	20.1397	624	0.028940	166	55	22.8013	645	0.036339	187
30	20.2021		0.029106		33 0	22.8658		0.036526	
		−624		−166			−645		−187
35	20.2645	624	0.029272	167	5	22.9303	646	0.036713	188
40	20.3269	62	0.029439	168	10	22.9949	647	0.036901	189
45	20.3894	625	0.029607	168	15	23.0596	647	0.037090	188
50	20.4519	626	0.029775	168	20	23.1243	648	0.037278	190
55	20.5145	626	0.029943	169	25	23.1891	649	0.037468	190
30 0	20.5771		0.030112		30	23.2540		0.037658	
		−627		−170			−649		−190
5	20.6398	627	0.030282	170	35	23.3189	649	0.037848	191
10	20.7025	628	0.030452	170	40	23.3838	650	0.038039	191
15	20.7653	628	0.030622	171	45	23.4488	651	0.038230	192
20	20.8281	629	0.030793	172	50	23.5139	651	0.038422	193
25	20.8910	629	0.030965	172	55	23.5790	652	0.038615	192
30	20.9539		0.031137		34 0	23.6442		0.038807	
		−630		−172			−653		−194
35	21.0169	630	0.031309	173	5	23.7095	653	0.039001	194
40	21.0799	631	0.031482	174	10	23.7748	653	0.039195	194
45	21.1430	631	0.031656	174	15	23.8401	654	0.039389	195
50	21.2061	632	0.031830	174	20	23.9055	655	0.039584	195
55	21.2693	633	0.032004	175	25	23.9710	656	0.039779	196
31 0	21.3326		0.032179		30	24.0366		0.039975	
		−632		−175			−656		−197
5	21.3958	634	0.032354	176	35	24.1022	656	0.040172	196
10	21.4592	634	0.032530	177	40	24.1678	657	0.040368	198
15	21.5226	634	0.032707	177	45	24.2335	658	0.040566	198
20	21.5860	635	0.032884	177	50	24.2993	658	0.040764	198
25	21.6495	635	0.033061	178	55	24.3651	659	0.040962	199
30	21.7130		0.033239		35 0	24.4310		0.041161	
		−636		−178			−660		−199

TABLE IV. *continued*.

Angle.		Mean Mot.	Diff.	Log. Diff.	Diff.	Angle.		Mean Mot.	Diff.	Log. Dift.	Diff.
°			—660		—199	°			—687		—221
35	5	24.4970	660	0.041360	200	38	35	27.3246	687	0.050195	221
	10	24.5630	661	0.041560	201		40	27.3933	688	0.050416	222
	15	24.6291	661	0.041761	201		45	27.4621	689	0.050638	223
	20	24.6952	662	0.041962	201		50	27.5310	690	0.050861	222
	25	24.7614	663	0.042163	202		55	27.6000	691	0.051083	224
	30	24.8277		0.042365		39	0	27.6691		0.051307	
			—663		—202				—691		—224
	35	24.8940	664	0.042567	203		5	27.7382	692	0.051531	224
	40	24.9604	664	0.042770	204		10	27.8074	692	0.051755	225
	45	25.0268	665	0.042974	204		15	27.8766	693	0.051980	226
	50	25.0933	666	0.043178	204		20	27.9459	694	0.052206	226
	55	25.1599	666	0.043382	205		25	28.0153	695	0.052432	226
36	0	25.2265		0.043587			30	28.0848		0.052658	
			—667		—206				—696		—227
	5	25.2932	668	0.043793	206		35	28.1544	696	0.052885	228
	10	25.3600	668	0.043999	206		40	28.2240	697	0.053113	228
	15	25.4268	669	0.044205	207		45	28.2937	697	0.053341	229
	20	25.4937	670	0.044412	208		50	28.3634	699	0.053570	229
	25	25.5607	670	0.044620	208		55	28.4333	699	0.053799	229
	30	25.6277		0.044828		40	0	28.5032		0.054028	
			—671		—209				—700		—231
	35	25.6948	671	0.045037	209		5	28.5732	700	0.054259	230
	40	25.7619	672	0.045246	209		10	28.6432	702	0.054489	231
	45	25.8291	673	0.045455	210		15	28.7134	702	0.054720	232
	50	25.8964	673	0.045665	211		20	28.7836	703	0.054952	232
	55	25.9637	674	0.045876	211		25	28.8539	703	0.055184	233
37	0	26.0311		0.046087			30	28.9242		0.055417	
			—675		—211				—704		—233
	5	26.0986	675	0.046298	213		35	28.9946	706	0.055650	234
	10	26.1661	676	0.046511	212		40	29.0652	706	0.055884	235
	15	26.2337	677	0.046723	213		45	29.1358	706	0.056119	234
	20	26.3014	677	0.046936	214		50	29.2064	708	0.056353	236
	25	26.3691	678	0.047150	214		55	29.2772	708	0.056589	236
	30	26.4369		0.047364		41	0	29.3480		0.056825	
			—679		—215				—709		—236
	35	26.5048	679	0.047579	215		5	29.4189	710	0.057061	237
	40	26.5727	680	0.047794	216		10	29.4899	710	0.057298	238
	45	26.6407	681	0.048010	216		15	29.5609	711	0.057536	238
	50	26.7088	682	0.048226	217		20	29.6320	712	0.057774	238
	55	26.7770	682	0.048443	217		25	29.7032	713	0.058012	239
38	0	26.8452		0.048660			30	29.7745		0.058251	
			—682		—218				—714		—240
	5	26.9134	684	0.048878	218		35	29.8459	714	0.058491	240
	10	26.9818	684	0.049096	218		40	29.9173	715	0.058731	240
	15	27.0502	685	0.049315	219		45	29.9888	716	0.058971	242
	20	27.1187	685	0.049534	219		50	30.0604	717	0.059213	241
	25	27.1872	687	0.049754	220		55	30.1321	718	0.059454	243
	30	27.2559		0.049974		42	0	30.2039		0.059697	
			—687		—221				—718		—242

TABLE IV. *continued.*

Angle.	Mean Mot.	Diff.	Log. Dift.	Diff.		Angle.	Log. Mean Mot.	Diff.	Log. Dift.	Diff.
° ′		−718		−242		° ′		−983		−265
42 5	30.2757	719	0.059939	241		45 35	1.523343	982	0.070614	266
10	30.3476	720	0.060183	243		40	1.524325	981	0.070880	266
15	30.4196	721	0.060426	245		45	1.525306	980	0.071146	267
20	30.4917	722	0.060671	244		50	1.526286	979	0.071413	267
25	30.5639	722	0.060915	246		55	1.527265	978	0.071680	268
30	30.6361		0.061161	246		46 0	1.528243		0.071948	
		−724		−246				−978		−268
35	30.7085	724	0.061407	246		5	1.529221	976	0.072216	269
40	30.7809	725	0.061653	247		10	1.530197	975	0.072485	270
45	30.8534	725	0.061900	248		15	1.531172	974	0.072755	270
50	30.9259	727	0.062148	248		20	1.532146	973	0.073025	270
55	30.9986	727	0.062396	248		25	1.533119	972	0.073295	271
43 0	31.0713		0.062644			30	1.534091		0.073566	
		−728		−249				−972		−272
5	31.1441	729	0.062893	250		35	1.535063	970	0.073838	272
10	31.2170	730	0.063143	250		40	1.536033	969	0.074110	273
15	31.2900	731	0.063393	251		45	1.537002	969	0.074383	273
20	31.3631	732	0.063644	251		50	1.537971	967	0.074656	274
25	31.4363	732	0.063895	252		55	1.538938	967	0.074930	274
30	31.5095		0.064147			47 0	1.539905		0.075204	
		−734		−252				−965		−275
35	31.5829	734	0.064399	253		5	1.540870	965	0.075479	276
40	31.6563	735	0.064652	253		10	1.541835	964	0.075755	276
45	31.7298	735	0.064905	254		15	1.542799	963	0.076031	276
50	31.8033	737	0.065159	254		20	1.543762	962	0.076307	278
55	31.8770	738	0.065413	255		25	1.544724	961	0.076585	277
44 0	31.9508		0.065668			30	1.545685		0.076862	
		−738		−256				−960		−278
5	32.0246	740	0.065924	256		35	1.546645	959	0.077140	279
10	32.0986	740	0.066180	256		40	1.547604	959	0.077419	279
15	32.1726	741	0.066436	257		45	1.548563	957	0.077698	280
20	32.2467	742	0.066693	258		50	1.549520	957	0.077978	281
25	32.3209	743	0.066951	258		55	1.550477	956	0.078259	281
30	32.3952		0.067209			48 0	1.551433		0.078540	
		−743		−259				−955		−281
35	32.4695	745	0.067468	259		5	1.552388	954	0.078821	282
40	32.5440	745	0.067727	260		10	1.553342	953	0.079103	283
45	32.6185	747	0.067987	260		15	1.554295	952	0.079386	283
50	32.6932	747	0.068247	261		20	1.555247	952	0.079669	284
55	32.7679	718	0.068508	261		25	1.556199	950	0.079953	284
45 0	32.8427		0.068769			30	1.557149		0.080237	
				−262				−950		−285
* 5	1.517428	989	0.069031	263		35	1.558099	949	0.080522	285
10	1.518417	987	0.069294	2?3		40	1.559048	948	0.080807	286
15	1.519404	986	0.069557	263		45	1.559996	948	0.081093	287
20	1.520390	986	0.069820	264		50	1.560944	946	0.081380	287
25	1.521376	984	0.070084	265		55	1.561890	946	0.081667	287
30	1.522360		0.070349			49 0	1.562836		0.081954	
		−983		−265				−945		−288

* Here the Second Column begins to be the Logarithm of the Mean Motion.

TABLE IV. *continued.*

Angle.		Log. Mean Mot.	Diff.	Log. Diff.	Diff.	Angle.		Log. Mean Mot.	Diff.	Log. Diff.	Diff.
°			—915		—288	°			—915		—312
49	5	1.563781		0.082242		52	35	1.602803		0.094850	
	10	1.564725	944	0.082531	289		40	1.603718	915	0.095162	312
	15	1.565668	943	0.082820	289		45	1.604632	914	0.095475	313
	20	1.566611	943	0.083110	290		50	1.605545	913	0.095789	314
	25	1.567553	942	0.083400	290		55	1.606458	913	0.096103	314
	30	1.568494	941	0.083691	291	53	0	1.607370	912	0.096418	315
			—940		—292				—912		—315
	35	1.569434	939	0.083983	292		5	1.608282	911	0.096733	315
	40	1.570373	939	0.084275	292		10	1.609193	911	0.097049	316
	45	1.571312	938	0.084567	292		15	1.610104	911	0.097365	316
	50	1.572250	937	0.084861	291		20	1.611014	910	0.097682	317
	55	1.573187	936	0.085154	293		25	1.611923	909	0.098000	318
50	0	1.574123		0.085449	295		30	1.612832	909	0.098318	318
			—936		—294				—908		—318
	5	1.575059	935	0.085743	296		35	1.613740	907	0.098636	319
	10	1.575994	934	0.086039	296		40	1.614647	907	0.098955	320
	15	1.576928	934	0.086335	296		45	1.615554	907	0.099275	321
	20	1.577862	932	0.086631	296		50	1.616461	906	0.099596	321
	25	1.578794	932	0.086928	297		55	1.617367	905	0.099917	321
	30	1.579726	932	0.087226	298	54	0	1.618272		0.100238	
			—932		—298				—905		—322
	35	1.580658	930	0.087524	299		5	1.619177	904	0.100560	323
	40	1.581588	930	0.087823	299		10	1.620081	904	0.100883	323
	45	1.582518	929	0.088122	300		15	1.620985	903	0.101206	324
	50	1.583447	928	0.088422	301		20	1.621888	903	0.101530	325
	55	1.584375	928	0.088723	301		25	1.622791	903	0.101855	325
51	0	1.585303		0.089024			30	1.623694		0.102180	
			—927		—301				—901		—325
	5	1.586230	926	0.089325	302		35	1.624595	901	0.102505	327
	10	1.587156	926	0.089627	303		40	1.625496	901	0.102832	326
	15	1.588082	925	0.089930	303		45	1.626397	900	0.103158	328
	20	1.589007	924	0.090233	304		50	1.627297	900	0.103486	328
	25	1.589931	924	0.090537	304		55	1.628197	899	0.103814	328
	30	1.590855		0.090841		55	0	1.629096		0.104142	
			—923		—305				—899		—329
	35	1.591778	922	0.091146	306		5	1.629995	898	0.104471	330
	40	1.592700	922	0.091452	306		10	1.630893	897	0.104801	330
	45	1.593622	921	0.091758	307		15	1.631790	898	0.105131	331
	50	1.594543	920	0.092065	307		20	1.632688	896	0.105462	332
	55	1.595463	920	0.092372	308		25	1.633584	896	0.105794	332
52	0	1.596383		0.092680			30	1.634480		0.106126	
			—919		—308				—896		—332
	5	1.597302	918	0.092988	309		35	1.635376	895	0.106458	333
	10	1.598220	918	0.093297	309		40	1.636271	895	0.106791	334
	15	1.599138	917	0.093606	309		45	1.637166	894	0.107125	335
	20	1.600055	917	0.093916	310		50	1.638060	894	0.107460	335
	25	1.600972	916	0.094227	311		55	1.638954	894	0.107795	335
	30	1.601888		0.094538	311	56	0	1.639848		0.108130	
			—915		—312				—892		—336

3 P

TABLE IV. *continued.*

Angle.	Log. Mean Mot.	Diff.	Log. Diff.	Diff.	Angle.	Log. Mean Mot.	Diff.	Log. Diff.	Diff.
		—892		—336			—877		—361
56° 5′	1.640740	893	0.108466	337	59° 35′	1.677878	877	0.123123	362
10	1.641633	892	0.108803	337	40	1.678755	876	0.123485	362
15	1.642525	892	0.109140	338	45	1.679631	876	0.123847	364
20	1.643417	891	0.109478	339	50	1.680507	876	0.124211	363
25	1.644308	891	0.109817	339	55	1.681383	875	0.124574	365
30	1.645199		0.110156		60 0	1.682258		0.124939	
		—890		—340			—875		—365
35	1.646089	889	0.110496	340	5	1.683133	875	0.125304	365
40	1.646978	890	0.110836	341	10	1.684008	875	0.125669	367
45	1.647868	889	0.111177	341	15	1.684883	874	0.126036	366
50	1.648757	888	0.111518	342	20	1.685757	874	0.126402	368
55	1.649645	889	0.111860	343	25	1.686631	873	0.126770	368
57 0	1.650534		0.112203		30	1.687504		0.127138	
		—887		—343			—874		—369
5	1.651421	888	0.112546	344	35	1.688378	873	0.127507	369
10	1.652309	887	0.112890	345	40	1.689251	873	0.127876	370
15	1.653196	886	0.113235	345	45	1.690124	873	0.128246	370
20	1.654082	886	0.113580	345	50	1.690997	872	0.128616	371
25	1.654968	886	0.113925	346	55	1.691869	872	0.128987	372
30	1.655854		0.114271		61 0	1.692741		0.129359	
		—885		—347			—872		—373
35	1.656739	885	0.114618	348	5	1.693613	871	0.129732	373
40	1.657624	885	0.114966	348	10	1.694484	871	0.130105	373
45	1.658509	884	0.115314	348	15	1.695355	871	0.130478	374
50	1.659393	884	0.115662	350	20	1.696226	871	0.130852	375
55	1.660277	883	0.116012	350	25	1.697097	871	0.131227	376
58 0	1.661160		0.116362		30	1.697968		0.131603	
		—883		—350			—870		—376
5	1.662043	883	0.116712	351	35	1.698838	870	0.131979	377
10	1.662926	882	0.117063	352	40	1.699708	870	0.132356	377
15	1.663808	882	0.117415	352	45	1.700578	870	0.132733	378
20	1.664690	882	0.117767	353	50	1.701448	869	0.133111	379
25	1.665572	881	0.118120	353	55	1.702317	869	0.133490	379
30	1.666453		0.118473		62 0	1.703186		0.133869	
		—881		—354			—869		—380
35	1.667334	880	0.118827	355	5	1.704055	869	0.134249	380
40	1.668214	881	0.119182	355	10	1.704924	869	0.134629	381
45	1.669095	880	0.119537	356	15	1.705793	868	0.135010	382
50	1.669975	879	0.119893	356	20	1.706661	868	0.135392	382
55	1.670854	879	0.120249	357	25	1.707529	868	0.135774	383
59 0	1.671733		0.120606		30	1.708397		0.136157	
		—879		—358			—868		—384
5	1.672612	878	0.120964	358	35	1.709265	868	0.136541	384
10	1.673490	879	0.121322	359	40	1.710133	867	0.136925	385
15	1.674369	878	0.121681	360	45	1.711000	867	0.137310	386
20	1.675247	877	0.122041	360	50	1.711867	867	0.137696	386
25	1.676124	877	0.122401	361	55	1.712734	867	0.138082	386
30	1.677001		0.122762		63 0	1.713601		0.138468	
		—877		—361			—866		—388

TABLE IV. *continued.*

Left section:

Angle.		Log. Mean Mot.	Diff.	Log. Dift.	Diff.
°	′		—866		—388
63	5	1.714467	867	0.138856	388
	10	1.715334	866	0.139244	388
	15	1.716200	866	0.139633	389
	20	1.717066	856	0.140022	389
	25	1.717932	865	0.140412	390
	30	1.718797		0.140802	390
			—866		—391
	35	1.719663	865	0.141193	
	40	1.720528	865	0.141585	392
	45	1.721393	855	0.141978	393
	50	1.722255	865	0.142371	393
	55	1.723123	865	0.142765	394
64	0	1.723988		0.143159	394
			—865		—395
	5	1.724853	864	0.143554	396
	10	1.725717	864	0.143950	396
	15	1.726581	865	0.144346	396
	20	1.727446	864	0.144743	397
	25	1.728310	863	0.145141	398
	30	1.729173		0.145539	398
			—864		—398
	35	1.730037	864	0.145938	399
	40	1.730901	863	0.146337	400
	45	1.731764	864	0.146737	401
	50	1.732628	863	0.147138	402
	55	1.733491	863	0.147540	402
65	0	1.734354		0.147942	402
			—863		—402
	5	1.735217	863	0.148344	404
	10	1.736080	863	0.148748	404
	15	1.736943	862	0.149152	404
	20	1.737805	863	0.149556	406
	25	1.738668	862	0.149962	406
	30	1.739530		0.150368	406
			—863		—406
	35	1.740393	862	0.150774	408
	40	1.741255	862	0.151182	408
	45	1.742117	862	0.151590	4c8
	50	1.742979	862	0.151998	409
	55	1.743841	862	0.152407	410
66	0	1.744703		0.152817	411
			—862		—411
	5	1.745565	862	0.153228	411
	10	1.746427	861	0.153639	412
	15	1.747288	862	0.154051	412
	20	1.748150	862	0.154463	413
	25	1.749012	861	0.154876	414
	30	1.749873		0.155290	414
			—861		—415

Right section:

Angle.		Log. Mean Mot.	Diff.	Log. Dift.	Diff.
°	′		—861		—415
66	35	1.750734	862	0.155705	415
	40	1.751596	861	0.156120	415
	45	1.752457	861	0.156536	416
	50	1.753318	861	0.156952	416
	55	1.754179	851	0.157369	417
67	0	1.755040		0.157787	418
			—862		—418
	5	1.755902	861	0.158205	419
	10	1.756763	860	0.158624	419
	15	1.757623	861	0.159044	420
	20	1.758484	861	0.159465	421
	25	1.759345	861	0.159886	421
	30	1.760206		0.160307	421
			—861		—423
	35	1.761067	861	0.160730	423
	40	1.761928	860	0.161153	423
	45	1.762788	861	0.161576	425
	50	1.763649	861	0.162001	425
	55	1.764510	861	0.162426	426
68	0	1.765371		0.162852	426
			—860		—426
	5	1.766231	861	0.163278	427
	10	1.767092	861	0.163705	428
	15	1.767953	860	0.164133	428
	20	1.768813	861	0.164561	429
	25	1.769674	861	0.164990	430
	30	1.770535		0.165420	430
			—860		—430
	35	1.771395	861	0.165850	431
	40	1.772256	860	0.166281	432
	45	1.773116	861	0.166713	433
	50	1.773977	861	0.167146	433
	55	1.774838	860	0.167579	434
69	0	1.775698		0.168013	434
			—861		—434
	5	1.776559	861	0.168447	435
	10	1.777420	861	0.168882	436
	15	1.778281	860	0.169318	436
	20	1.779141	861	0.169754	438
	25	1.780002	861	0.170192	438
	30	1.780863		0.170630	438
			—861		—438
	35	1.781724	860	0.171068	439
	40	1.782584	861	0.171507	440
	45	1.783445	861	0.171947	441
	50	1.784306	861	0.172388	441
	55	1.785167	861	0.172829	442
70	0	1.786028		0.173271	443
			—861		—443

TABLE IV. *continued.*

Angle		Log. Mean Mot.	Diff.	Log. Diff.	Diff.	Angle		Log. Mean Mot.	Diff.	Log. Dift.	Diff.
°	′		−861		−443	°	′		−865		172
70	5	1.786889	862	0.173714	443	73	35	1.823131	866	0.192932	473
	10	1.787751	861	0.174157	443		40	1.823997	865	0.193405	473
	15	1.788612	861	0.174601	444		45	1.824862	866	0.193878	474
	20	1.789473	861	0.175046	445		50	1.825728	866	0.194352	475
	25	1.790334	861	0.175491	445		55	1.826594	866	0.194827	476
	30	1.791195		0.175937	446	74	0	1.827460		0.195303	476
			−862		−447				−867		−476
	35	1.792057	861	0.176384	447		5	1.828327	866	0.195779	477
	40	1.792918	862	0.176831	448		10	1.829193	867	0.196256	478
	45	1.793780	862	0.177279	449		15	1.830060	867	0.196734	479
	50	1.794642	861	0.177728	450		20	1.830927	867	0.197213	479
	55	1.795503	862	0.178178	450		25	1.831794	867	0.197692	480
71	0	1.796365		0.178628			30	1.832661		0.198172	
			−862		−451				−868		−480
	5	1.797227	862	0.179079	451		35	1.833529	867	0.198652	482
	10	1.798089	862	0.179530	453		40	1.834396	868	0.199134	482
	15	1.798951	862	0.179983	453		45	1.835264	868	0.199616	483
	20	1.799813	862	0.180436	453		50	1.836132	868	0.200099	483
	25	1.800675	862	0.180889	455		55	1.837000	869	0.200582	485
	30	1.801537		0.181344		75	0	1.837869		0.201067	
			−863		−455				−868		−485
	35	1.802400	862	0.181799	456		5	1.838737	869	0.201552	486
	40	1.803262	862	0.182255	456		10	1.839606	869	0.202038	486
	45	1.804124	863	0.182711	457		15	1.840475	869	0.202524	487
	50	1.804987	863	0.183168	458		20	1.841344	870	0.203011	488
	55	1.805850	863	0.183626	459		25	1.842214	869	0.203499	489
72	0	1.806713		0.184085			30	1.843083		0.203988	
			−863		−459				−870		−489
	5	1.807576	863	0.184544	460		35	1.843953	870	0.204477	491
	10	1.808439	863	0.185004	461		40	1.844823	871	0.204968	491
	15	1.809302	863	0.185465	461		45	1.845694	870	0.205459	491
	20	1.810165	864	0.185926	462		50	1.846564	871	0.205950	493
	25	1.811029	863	0.186388	463		55	1.847435	871	0.206443	493
	30	1.811892		0.186851		76	0	1.848306		0.206936	
			−864		−463				−871		−494
	35	1.812756	864	0.187314	465		5	1.849177	872	0.207430	494
	40	1.813620	864	0.187779	465		10	1.850049	871	0.207924	496
	45	1.814484	864	0.188244	465		15	1.850920	872	0.208420	496
	50	1.815348	864	0.188709	467		20	1.851792	872	0.208916	496
	55	1.816212	865	0.189176	467		25	1.852664	873	0.209412	498
73	0	1.817077		0.189643			30	1.853537		0.209910	
			−864		−467				−873		−498
	5	1.817941	864	0.190110	469		35	1.854410	873	0.210408	499
	10	1.818805	865	0.190579	469		40	1.855283	873	0.210907	500
	15	1.819670	865	0.191048	470		45	1.856156	873	0.211407	501
	20	1.820535	865	0.191518	470		50	1.857029	874	0.211908	501
	25	1.821400	866	0.191988	472		55	1.857903	874	0.212409	502
	30	1.822266		0.192460		77	0	1.858777		0.212911	
			−865		−472				−874		−503

TABLE IV. continued.

Angle.		Log. Mean Mot.	Diff.	Log. Dift.	Diff.	Angle.		Log. Mean Mot.	Diff.	Log. Dift.	Diff.
°	'		—874		—503	°	'		—887		—535
77	5	1.859651	875	0.213414	504	80	35	1.896640	888	0.235221	536
	10	1.860526	874	0.213918	504		40	1.897528	889	0.235757	537
	15	1.861400	875	0.214422	505		45	1.898417	888	0.236294	538
	20	1.862275	876	0.214927	506		50	1.899305	889	0.236832	538
	25	1.863151	875	0.215433	506		55	1.900194	890	0.237370	539
	30	1.864026		0.215939		81	0	1.901084		0.237909	
			—876		—5c8				—890		—540
	35	1.864902	876	0.216447	508		5	1.901974	890	0.238449	541
	40	1.865778	877	0.216955	508		10	1.902864	891	0.238990	541
	45	1.866655	877	0.217464	509		15	1.903755	891	0.239531	542
	50	1.867532	876	0.217973	509		20	1.904646	891	0.240073	543
	55	1.868408	878	0.218484	511		25	1.905537	892	0.240616	544
78	0	1.869286		0.218995	511		30	1.906429		0.241160	
			—877		—512				—892		—545
	5	1.870163	878	0.219507	512		35	1.907321	893	0.241705	545
	10	1.871041	878	0.220019	514		40	1.908214	893	0.242250	547
	15	1.871919	879	0.220533	514		45	1.909107	894	0.242797	547
	20	1.872798	879	0.221047	515		50	1.910001	894	0.243344	548
	25	1.873677	879	0.221562	516		55	1.910895	894	0.243892	548
	30	1.874556		0.222078		82	0	1.911789		0.244440	
			—879		—516				—895		—550
	35	1.875435	880	0.222594	517		5	1.912684	895	0.244990	550
	40	1.876315	880	0.223111	518		10	1.913579	896	0.245540	551
	45	1.877195	880	0.223629	519		15	1.914475	896	0.246091	552
	50	1.878075	881	0.224148	520		20	1.915371	896	0.246643	553
	55	1.878956	881	0.224668	520		25	1.916267	897	0.247196	553
79	0	1.879837		0.225188			30	1.917164		0.247749	
			—881		—521				—897		—555
	5	1.880718	882	0.225709	522		35	1.918061	898	0.248304	555
	10	1.881600	882	0.226231	522		40	1.918959	898	0.248859	556
	15	1.882482	882	0.226753	524		45	1.919857	899	0.249415	557
	20	1.883364	883	0.227277	524		50	1.920756	899	0.249972	557
	25	1.884247	883	0.227801	525		55	1.921655	900	0.250529	559
	30	1.885130		0.228326		83	0	1.922555		0.251088	
			—883		—526				—900		—559
	35	1.886013	884	0.228852	526		5	1.923455	900	0.251647	560
	40	1.886897	884	0.229378	528		10	1.924355	901	0.252207	561
	45	1.887781	884	0.229906	528		15	1.925256	901	0.252768	562
	50	1.888665	885	0.230434	528		20	1.926157	902	0.253330	562
	55	1.889550	885	0.230962	530		25	1.927059	903	0.253892	564
80	0	1.890435		0.231492			30	1.927962		0.254456	
			—885		—530				—903		—564
	5	1.891320	886	0.232022	532		35	1.928865	903	0.255020	565
	10	1.892206	886	0.232554	532		40	1.929768	903	0.255585	566
	15	1.893092	887	0.233086	532		45	1.930671	904	0.256151	566
	20	1.893979	887	0.233618	534		50	1.931575	905	0.256717	568
	25	1.894866	887	0.234152	534		55	1.932480	905	0.257285	568
	30	1.895753		0.234686		84	0	1.933385		0.257853	
			—887		—535				—906		—569

TABLE IV. continued.

Left section

Angle (°)	Angle (′)	Log. Mean Mot.	Diff.	Log. Diff.	Diff.
			—906		—563
84	5	1.934291		0.258422	
			906		570
	10	1.935197		0.258992	
			907		571
	15	1.936104		0.259563	
			907		572
	20	1.937011		0.260135	
			907		572
	25	1.937918		0.260707	
			908		574
	30	1.938826		0.261281	
			—909		—574
	35	1.939735		0.261855	
			909		575
	40	1.940644		0.262430	
			910		576
	45	1.941554		0.263006	
			910		576
	50	1.942464		0.263582	
			910		578
	55	1.943374		0.264160	
			911		578
85	0	1.944285		0.264738	
			—912		—579
	5	1.945197		0.265317	
			912		580
	10	1.946109		0.265897	
			913		581
	15	1.947022		0.266478	
			913		582
	20	1.947935		0.267060	
			914		583
	25	1.948849		0.267643	
			914		583
	30	1.949763		0.268226	
			—915		—585
	35	1.950678		0.268811	
			915		585
	40	1.951593		0.269396	
			916		586
	45	1.952509		0.269982	
			917		587
	50	1.953426		0.270569	
			917		587
	55	1.954343		0.271156	
			917		589
86	0	1.955260		0.271745	
			—918		—590
	5	1.956178		0.272335	
			919		590
	10	1.957097		0.272925	
			919		591
	15	1.958016		0.273516	
			920		592
	20	1.958936		0.274108	
			920		593
	25	1.959856		0.274701	
			921		594
	30	1.960777		0.275295	
			—921		—594
	35	1.961698		0.275889	
			922		596
	40	1.962620		0.276485	
			923		596
	45	1.963543		0.277081	
			923		597
	50	1.964466		0.277678	
			924		599
	55	1.965390		0.278277	
			924		599
87	0	1.966314		0.278876	
			—925		—599
	5	1.967239		0.279475	
			925		601
	10	1.968164		0.280072	
			926		602
	15	1.969090		0.280678	
			927		602
	20	1.970017		0.281280	
			927		604
	25	1.970944		0.281884	
			928		604
	30	1.971872		0.282488	
			—929		—605

Right section

Angle (°)	Angle (′)	Log. Mean Mot.	Diff.	Log. Diff.	Diff.
			—929		—605
87	35	1.972801		0.283093	
			929		606
	40	1.973730		0.283699	
			929		607
	45	1.974659		0.284306	
			930		608
	50	1.975589		0.284914	
			931		608
	55	1.976520		0.285522	
			932		610
88	0	1.977452		0.286132	
			—932		—610
	5	1.978384		0.286742	
			933		612
	10	1.979317		0.287354	
			933		612
	15	1.980250		0.287966	
			934		613
	20	1.981184		0.288579	
			935		614
	25	1.982119		0.289193	
			935		615
	30	1.983054		0.289808	
			—936		—616
	35	1.983990		0.290424	
			936		616
	40	1.984926		0.291040	
			938		618
	45	1.985864		0.291658	
			937		618
	50	1.986801		0.292276	
			939		620
	55	1.987740		0.292896	
			939		620
89	0	1.988679		0.293516	
			—940		—621
	5	1.989619		0.294137	
			940		622
	10	1.990559		0.294759	
			941		623
	15	1.991500		0.295382	
			942		624
	20	1.992442		0.296006	
			942		625
	25	1.993384		0.296631	
			943		626
	30	1.994327		0.297257	
			—944		—626
	35	1.995271		0.297883	
			945		628
	40	1.996216		0.298511	
			945		628
	45	1.997161		0.299139	
			946		630
	50	1.998107		0.299769	
			946		630
	55	1.999053		0.300399	
			947		631
90	0	2.000000		0.301030	
			—948		—632
	5	2.000948		0.301662	
			949		633
	10	2.001897		0.302295	
			949		634
	15	2.002846		0.302929	
			950		635
	20	2.003796		0.303564	
			950		636
	25	2.004746		0.304200	
			951		637
	30	2.005697		0.304837	
			—952		—637
	35	2.006649		0.305474	
			953		639
	40	2.007602		0.306113	
			954		639
	45	2.008556		0.306752	
			954		641
	50	2.009510		0.307393	
			955		641
	55	2.010465		0.308034	
			955		642
91	0	2.011420		0.308676	
			—957		—644

TABLE IV. continued.

Angle.	Log. Mean Mot.	Diff.	Log. Diff.	Diff.	Angle.	Log. Mean Mot.	Diff.	Log. Diff.	Diff.
		—957		—644			—989		—581
91° 5	2.012377	957	0.309320	644	94° 35	2.053239	991	0.337199	685
10	2.013334	957	0.309964	644	40	2.054230	991	0.337884	686
15	2.014292	958	0.310609	645	45	2.055221	992	0.338570	687
20	2.015250	958	0.311255	646	50	2.056213	993	0.339257	687
25	2.016209	959	0.311902	647	55	2.057206	994	0.339944	689
30	2.017169	960	0.312550	648	95 0	2.058200		0.340633	
		—961		—649			—995		—690
35	2.018130	961	0.313199	650	5	2.059195	996	0.341323	690
40	2.019091	963	0.313849	650	10	2.060191	997	0.342014	691
45	2.020054	963	0.314499	652	15	2.061188	997	0.342706	692
50	2.021017	964	0.315151	653	20	2.062185	999	0.343399	693
55	2.021981	964	0.315804	653	25	2.063184	999	0.344093	694
92 0	2.022945		0.316457		30	2.064183		0.344787	694
		—965		—655			—1000		—696
5	2.023910	966	0.317112	656	35	2.065183	1001	0.345483	697
10	2.024876	967	0.317768	656	40	2.066184	1002	0.346136	698
15	2.025843	968	0.318424	657	45	2.067186	1003	0.346878	699
20	2.026811	968	0.319081	659	50	2.068189	1004	0.347577	700
25	2.027779	969	0.319740	659	55	2.069193	1005	0.348277	701
30	2.028748		0.320399		96 0	2.070198		0.348978	
		—970		—661			—1006		—702
35	2.029718	971	0.321060	661	5	2.071204	1007	0.349680	703
40	2.030689	971	0.321721	662	10	2.072211	1007	0.350383	704
45	2.031660	973	0.322383	663	15	2.073218	1008	0.351087	706
50	2.032633	973	0.323046	664	20	2.074226	1010	0.351793	706
55	2.033606	974	0.323710	666	25	2.075236	1010	0.352499	707
93 0	2.034580		0.324376		30	2.076246		0.353206	
		—974		—666			—1011		—708
5	2.035554	976	0.325042	667	35	2.077257	1013	0.353914	709
10	2.036530	976	0.325709	668	40	2.078270	1013	0.354623	711
15	2.037506	977	0.326377	669	45	2.079283	1014	0.355334	711
20	2.038483	978	0.327046	670	50	2.080297	1015	0.356045	712
25	2.039461	979	0.327716	671	55	2.081312	1016	0.356757	714
30	2.040440		0.328387		97 0	2.082328		0.357471	
		—979		—672			—1017		—714
35	2.041419	981	0.329059	673	5	2.083345	1018	0.358185	716
40	2.042400	981	0.329732	674	10	2.084363	1019	0.358901	716
45	2.043381	982	0.330406	675	15	2.085382	1020	0.359617	718
50	2.044363	983	0.331081	676	20	2.086402	1021	0.360335	719
55	2.045346	984	0.331757	676	25	2.087423	1022	0.361054	719
94 0	2.046330		0.332433		30	2.088445		0.361773	
		—984		—678			—1023		—721
5	2.047314	985	0.333111	679	35	2.089468	1024	0.362494	722
10	2.048299	987	0.333790	680	40	2.090492	1025	0.363216	723
15	2.049286	987	0.334470	681	45	2.091517	1025	0.363939	724
20	2.050273	988	0.335151	682	50	2.092542	1027	0.364663	725
25	2.051261	989	0.335833	682	55	2.093569	1028	0.365388	726
30	2.052250		0.336515		98 0	2.094597		0.366114	
		—989		—684			—1029		—727

TABLE IV. *continued.*

Angle.	Log. Mean Mot.	Diff.	Log. Dist.	Diff.
		—1029		—727
98 5	2.095626	1030	0.366841	729
10	2.096656	1031	0.367570	729
15	2.097687	1032	0.368299	730
20	2.098719	1033	0.369029	732
25	2.099752	1033	0.369761	732
30	2.100785		0.370493	
		—1035		—734
35	2.101820	1036	0.371227	735
40	2.102856	1037	0.371962	735
45	2.103893	1038	0.372697	737
50	2.104931	1040	0.373434	738
55	2.105971	1040	0.374172	739
99 0	2.107011		0.374911	
		—1041		—740
5	2.108052	1042	0.375651	741
10	2.109094	1044	0.376392	743
15	2.110138	1044	0.377135	743
20	2.111182	1045	0.377878	745
25	2.112227	1047	0.378623	745
30	2.113274		0.379368	
		—1048		—747
35	2.114322	1048	0.380115	748
40	2.115370	1050	0.380863	749
45	2.116420	1051	0.381612	750
50	2.117471	1052	0.382362	751
55	2.118523	1053	0.383113	752
100 0	2.119576		0.383865	
		—1054		—753
5	2.120630	1055	0.384618	755
10	2.121685	1057	0.385373	755
15	2.122742	1057	0.386128	757
20	2.123799	1058	0.386885	758
25	2.124857	1060	0.387643	759
30	2.125917		0.388402	
		—1061		—760
35	2.126978	1062	0.389162	761
40	2.128040	1063	0.389923	762
45	2.129103	1064	0.390685	764
50	2.130167	1065	0.391449	764
55	2.131232	1067	0.392213	766
101 0	2.132299		0.392979	
		—1068		—767
5	2.133367	1068	0.393746	768
10	2.134435	1070	0.394514	769
15	2.135505	1071	0.395283	770
20	2.136576	1072	0.396053	771
25	2.137648	1074	0.396824	773
30	2.138722		0.397597	
		—1074		—774

Angle.	Log. Mean Mot.	Diff.	Log. Dist.	Diff.
		—1074		—774
101 35	2.139796	1076	0.398371	775
40	2.140872	1077	0.399146	776
45	2.141949	1078	0.399922	777
50	2.143027	1079	0.400699	778
55	2.144106	1081	0.401477	779
102 0	2.145187		0.402256	
		—1081		—781
5	2.146268	1083	0.403037	782
10	2.147351	1084	0.403819	783
15	2.148435	1085	0.404602	784
20	2.149520	1087	0.405386	785
25	2.150607	1083	0.406171	786
30	2.151695		0.406957	
		—1088		—788
35	2.152783	1090	0.407745	789
40	2.153873	1092	0.408534	790
45	2.154965	1092	0.409324	791
50	2.156057	1094	0.410115	792
55	2.157151	1095	0.410907	794
103 0	2.158246		0.411701	
		—1096		—795
5	2.159342	1098	0.412496	796
10	2.160440	1099	0.413292	797
15	2.161539	1100	0.414089	798
20	2.162639	1101	0.414887	799
25	2.163740	1102	0.415686	801
30	2.164842		0.416487	
		—1104		—802
35	2.165946	1105	0.417289	803
40	2.167051	1107	0.418092	804
45	2.168158	1107	0.418896	805
50	2.169265	1109	0.419701	807
55	2.170374	1110	0.420508	808
104 0	2.171484		0.421316	
		—1112		—809
5	2.172596	1113	0.422125	810
10	2.173709	1114	0.422935	812
15	2.174823	1115	0.423747	813
20	2.175938	1117	0.424560	814
25	2.177055	1118	0.425374	815
30	2.178173		0.426189	
		—1119		—816
35	2.179292	1121	0.427005	818
40	2.180413	1122	0.427823	819
45	2.181535	1123	0.428642	820
50	2.182658	1125	0.429462	821
55	2.183783	1126	0.430283	823
105 0	2.184909		0.431106	
		—1127		—824

TABLE IV. *continued.*

Angle.	Log. Mean Mot.	Diff.	Log. Dist.	Diff.	Angle.	Log. Mean Mot	Diff.	Log. Dist.	Diff.
° ′		—1127		—824	° ′		—1188		—878
105 5	2.186036	1129	0.431930	825	108 35	2.234666	1190	0.467681	880
10	2.187165	1130	0.432755	826	40	2.235856	1192	0.468561	880
15	2.188295	1132	0.433581	827	45	2.237048	1193	0.469441	883
20	2.189427	1133	0.434408	829	50	2.238241	1194	0.470324	883
25	2.190560	1134	0.435237	830	55	2.239435	1196	0.471207	885
30	2.191694		0.436067		109 0	2.240631		0.472092	
		—1136		—832			—1198		—886
35	2.192830	1137	0.436899	832	5	2.241829	1199	0.472978	888
40	2.193967	1138	0.437731	834	10	2.243028	1201	0.473866	889
45	2.195105	1140	0.438565	835	15	2.244229	1203	0.474755	890
50	2.196245	1141	0.439400	836	20	2.245432	1204	0.475645	892
55	2.197386	1142	0.440236	838	25	2.246636	1206	0.476537	893
106 0	2.198528		0.441074		30	2.247842		0.477430	
		—1144		—839			—1207		—894
5	2.199672	1145	0.441913	840	35	2.249049	1209	0.478324	896
10	2.200817	1147	0.442753	841	40	2.250258	1210	0.479220	897
15	2.201964	1148	0.443594	843	45	2.251468	1212	0.480117	899
20	2.203112	1150	0.444437	844	50	2.252680	1214	0.481016	900
25	2.204262	1151	0.445281	845	55	2.253894	1216	0.481916	901
30	2.205413		0.446126		110 0	2.255110		0.482817	
		—1152		—847			—1217		—903
35	2.206565	1154	0.446973	848	5	2.256327	1219	0.483720	904
40	2.207719	1156	0.447821	849	10	2.257546	1220	0.484624	906
45	2.208875	1156	0.448670	850	15	2.258766	1222	0.485530	907
50	2.210031	1158	0.449520	852	20	2.259988	1224	0.486437	908
55	2.211189	1160	0.450372	853	25	2.261212	1225	0.487345	910
107 0	2.212349		0.451225		30	2.262437		0.488255	
		—1161		—854			—1227		—911
5	2.213510	1163	0.452079	856	35	2.263664	1229	0.489166	913
10	2.214673	1164	0.452935	857	40	2.264893	1231	0.490079	914
15	2.215837	1166	0.453792	858	45	2.266124	1232	0.490993	916
20	2.217003	1167	0.454650	859	50	2.267356	1234	0.491909	917
25	2.218170	1168	0.455509	861	55	2.268590	1236	0.492826	918
30	2.219338		0.456370		111 0	2.269826		0.493744	
		—1170		—862			—1237		—920
35	2.220508	1172	0.457232	864	5	2.271063	1239	0.494664	921
40	2.221680	1173	0.458096	864	10	2.272302	1241	0.495585	923
45	2.222853	1174	0.458960	866	15	2.273543	1242	0.496508	924
50	2.224027	1176	0.459826	868	20	2.274785	1244	0.497432	925
55	2.225203	1178	0.460694	869	25	2.276029	1246	0.498357	927
108 0	2.226381		0.461563		30	2.277275		0.499284	
		—1179		—870			—1248		—929
5	2.227560	1180	0.462433	871	35	2.278523	1249	0.500213	930
10	2.228740	1182	0.463304	873	40	2.279772	1251	0.501143	931
15	2.229922	1184	0.464177	874	45	2.281023	1253	0.502074	933
20	2.231106	1185	0.465051	875	50	2.282276	1255	0.503007	934
25	2.232291	1187	0.465926	877	55	2.283531	1257	0.503941	936
30	2.233478		0.466803		112 0	2.284788		0.504877	
		—1188		—878			—1258		—937

3 R

TABLE IV. *continued*.

Angle.	Log. Mean Mot.	Diff.	Log. Dift.	Diff.
° ′		−1258		−937
112 5	2.286046		0.505814	
		1260		939
10	2.287306		0.506753	
		1262		940
15	2.288568		0.507693	
		1264		941
20	2.289832		0.508634	
		1265		943
25	2.291097		0.509577	
		1267		945
30	2.292364		0.510522	
		−1270		−946
35	2.293634		0.511468	
		1271		948
40	2.294905		0.512416	
		1272		949
45	2.296177		0.513365	
		1275		950
50	2.297452		0.514315	
		1276		952
55	2.298728		0.515267	
		1279		954
113 0	2.300007		0.516221	
		−1280		−955
5	2.301287		0.517176	
		1282		957
10	2.302569		0.518133	
		1284		958
15	2.303853		0.519091	
		1286		959
20	2.305139		0.520050	
		1287		962
25	2.306426		0.521012	
		1290		962
30	2.307716		0.521974	
		−1291		−964
35	2.309007		0.522938	
		1293		966
40	2.310300		0.523904	
		1296		967
45	2.311596		0.524871	
		1297		969
50	2.312893		0.525840	
		1299		971
55	2.314192		0.526811	
		1301		971
114 0	2.315493		0.527782	
		−1303		−974
5	2.316796		0.528756	
		1305		975
10	2.318101		0.529731	
		1306		976
15	2.319407		0.530707	
		1309		979
20	2.320716		0.531686	
		1311		979
25	2.322027		0.532665	
		1312		981
30	2.323339		0.533646	
		−1315		−983
35	2.324654		0.534629	
		1316		985
40	2.325970		0.535614	
		1319		986
45	2.327289		0.536600	
		1320		987
50	2.328609		0.537587	
		1323		989
55	2.329932		0.538576	
		1325		991
115 0	2.331257		0.539567	
		−1326		−992
5	2.332583		0.540559	
		1329		994
10	2.333912		0.541553	
		1330		996
15	2.335242		0.542549	
		1333		997
20	2.336575		0.543546	
		1334		999
25	2.337909		0.544545	
		1337		1000
30	2.339246		0.545545	
		−1339		−1002

Angle.	Log. Mean Mot.	Diff.	Log. Dift.	Diff.
° ′		−1339		−1002
115 35	2.340585		0.546547	
		1340		1003
40	2.341925		0.547550	
		1343		1005
45	2.343268		0.548555	
		1345		1007
50	2.344613		0.549562	
		1347		1009
55	2.345960		0.550571	
		1349		1010
116 0	2.347309		0.551581	
		−1351		−1011
5	2.348660		0.552592	
		1354		1014
10	2.350014		0.553606	
		1355		1015
15	2.351369		0.554621	
		1357		1016
20	2.352726		0.555637	
		1360		1018
25	2.354086		0.556655	
		1362		1020
30	2.355448		0.557675	
		−1364		−1022
35	2.356812		0.558697	
		1365		1023
40	2.358177		0.559720	
		1368		1025
45	2.359545		0.560745	
		1371		1027
50	2.360916		0.561772	
		1372		1028
55	2.362288		0.562800	
		1375		1030
117 0	2.363663		0.563830	
		−1377		−1032
5	2.365040		0.564862	
		1378		1033
10	2.366418		0.565895	
		1381		1035
15	2.367799		0.566930	
		1384		1036
20	2.369183		0.567966	
		1385		1039
25	2.370568		0.569005	
		1388		1040
30	2.371956		0.570045	
		−1390		−1042
35	2.373346		0.571087	
		1392		1043
40	2.374738		0.572130	
		1394		1045
45	2.376132		0.573175	
		1397		1047
50	2.377529		0.574222	
		1399		1049
55	2.378928		0.575271	
		1401		1050
118 0	2.380329		0.576321	
		−1403		−1052
5	2.381732		0.577373	
		1406		1054
10	2.383138		0.578427	
		1408		1056
15	2.384546		0.579483	
		1410		1057
20	2.385956		0.580540	
		1413		1059
25	2.387369		0.581599	
		1415		1061
30	2.388784		0.582660	
		−1417		−1063
35	2.390201		0.583723	
		1419		1064
40	2.391620		0.584787	
		1422		1066
45	2.393042		0.585853	
		1424		1068
50	2.394466		0.586921	
		1426		1070
55	2.395892		0.587991	
		1429		1071
119 0	2.397321		0.589062	
		−1431		−1074

TABLE IV. *continued.*

Angle	Log. Mean Mot.	Diff.	Log. Dist.	Diff.	Angle	Log. Mean Mot.	Diff.	Log. Dist.	Diff.
119° 5′	2.398752	−1431	0.590136	−1074	122° 35′	2.461092	−1537	0.636882	−1152
10	2.400186	1434	0.591211	1075	40	2.462632	1540	0.638037	1155
15	2.401622	1436	0.592287	1076	45	2.464175	1543	0.639193	1156
20	2.403060	1438	0.593366	1079	50	2.465721	1546	0.640351	1158
25	2.404500	1440	0.594446	1080	55	2.467269	1548	0.641512	1161
30	2.405943	1443	0.595529	1083	123° 0′	2.468821	1552	0.642674	1162
35	2.407389	−1446	0.596613	−1084	5	2.470375	−1554	0.643839	−1165
40	2.408836	1447	0.597698	1085	10	2.471931	1556	0.645005	1166
45	2.410287	1451	0.598786	1088	15	2.473491	1560	0.646173	1168
50	2.411739	1452	0.599876	1090	20	2.475053	1562	0.647344	1171
55	2.413194	1455	0.600967	1091	25	2.476618	1565	0.648516	1172
120° 0′	2.414652	1458	0.602060	1093	30	2.478186	1568	0.649691	1175
5	2.416112	−1460	0.603155	−1095	35	2.479756	−1570	0.650867	−1176
10	2.417574	1462	0.604252	1097	40	2.481330	1574	0.652046	1179
15	2.419039	1465	0.605350	1098	45	2.482907	1577	0.653227	1181
20	2.420507	1468	0.606451	1101	50	2.484486	1579	0.654410	1183
25	2.421977	1470	0.607553	1102	55	2.486068	1582	0.655594	1184
30	2.423449	1472	0.608658	1105	124° 0′	2.487653	1585	0.656781	1187
35	2.424924	−1475	0.609764	−1106	5	2.489241	−1588	0.657970	−1189
40	2.426401	1477	0.610872	1108	10	2.490832	1591	0.659162	1192
45	2.427881	1480	0.611982	1110	15	2.492425	1593	0.660355	1193
50	2.429363	1482	0.613093	1111	20	2.494022	1597	0.661550	1195
55	2.430848	1485	0.614207	1114	25	2.495621	1599	0.662747	1197
121° 0′	2.432336	1488	0.615322	1115	30	2.497224	1603	0.663947	1200
5	2.433826	−1490	0.616440	−1118	35	2.498829	−1605	0.665149	−1202
10	2.435318	1492	0.617559	1119	40	2.500437	1608	0.666352	1203
15	2.436813	1495	0.618680	1121	45	2.502049	1612	0.667558	1206
20	2.438311	1498	0.619803	1123	50	2.503663	1614	0.668766	1208
25	2.439811	1500	0.620928	1125	55	2.505280	1617	0.669977	1211
30	2.441314	1503	0.622055	1127	125° 0′	2.506901	1621	0.671189	1212
35	2.442820	−1506	0.623184	−1129	5	2.508524	−1623	0.672403	−1214
40	2.444328	1508	0.624315	1131	10	2.510150	1626	0.673620	1217
45	2.445839	1511	0.625448	1133	15	2.511780	1630	0.674839	1219
50	2.447352	1513	0.626582	1134	20	2.513412	1632	0.676060	1221
55	2.448868	1516	0.627719	1137	25	2.515047	1635	0.677283	1223
122° 0′	2.450387	1519	0.628858	1139	30	2.516686	1639	0.678508	1225
5	2.451908	−1521	0.629998	−1140	35	2.518327	−1642	0.679736	−1228
10	2.453432	1524	0.631141	1143	40	2.519972	1646	0.680965	1229
15	2.454959	1527	0.632285	1144	45	2.521620	1648	0.682197	1232
20	2.456488	1529	0.633431	1146	50	2.523270	1650	0.683432	1235
25	2.458020	1532	0.634580	1149	55	2.524924	1651	0.684668	1236
30	2.459555	1535	0.635730	1150	126° 0′	2.526581	1657	0.685906	1238
		−1537		−1152			−1660		−241

TABLE IV. *continued.*

Angle.	Log. Mean Mot.	Diff.	Log. Dift.	Diff.	Angle.	Log. Mean Mot.	Diff.	Log. Dift.	Diff.
° ′		—1660		—1241	° ′		—1802		—1341
126 5	2.528241	1664	0.687147	1243	129 35	2.600952	1806	0.741363	1343
10	2.529905	1666	0.688390	1246	40	2.602758	1810	0.742706	1345
15	2.531571	1670	0.689636	1247	45	2.604568	1813	0.744051	1348
20	2.533241	1673	0.690883	1250	50	2.606381	1817	0.745399	1351
25	2.534914	1676	0.692133	1252	55	2.608198	1821	0.746750	1353
30	2.536590		0.693385		130 0	2.610019		0.748103	
		—1679		—1254			—1824		—1356
35	2.538269	1682	0.694639	1257	5	2.611843	1828	0.749459	1359
40	2.539951	1686	0.695896	1259	10	2.613671	1832	0.750818	1361
45	2.541637	1689	0.697155	1261	15	2.615503	1836	0.752179	1364
50	2.543326	1692	0.698416	1263	20	2.617339	1840	0.753543	1366
55	2.545018	1696	0.699679	1266	25	2.619179	1843	0.754909	1369
127 0	2.546714		0.700945		30	2.621022		0.756278	
		—1698		—1268			—1847		—1371
5	2.548412	1702	0.702213	1271	35	2.622869	1851	0.757649	1374
10	2.550114	1705	0.703484	1272	40	2.624720	1855	0.759023	1377
15	2.551819	1709	0.704756	1275	45	2.626575	1859	0.760400	1379
20	2.553528	1712	0.706031	1278	50	2.628434	1862	0.761779	1382
25	2.555240	1715	0.707309	1279	55	2.630296	1866	0.763161	1385
30	2.556955		0.708588		131 0	2.632162		0.764546	
		—1719		—1282			—1870		—1387
35	2.558674	1721	0.709870	1285	5	2.634032	1875	0.765933	1391
40	2.560395	1726	0.711155	1287	10	2.635907	1878	0.767324	1392
45	2.562121	1728	0.712442	1289	15	2.637785	1882	0.768716	1396
50	2.563849	1732	0.713731	1291	20	2.639667	1886	0.770112	1398
55	2.565581	1735	0.715022	1294	25	2.641553	1890	0.771510	1401
128 0	2.567316		0.716316		30	2.643443		0.772911	
		—1739		—1296			—1893		—1403
5	2.569055	1743	0.717612	1299	35	2.645336	1898	0.774314	1407
10	2.570798	1746	0.718911	1301	40	2.647234	1902	0.775721	1409
15	2.572544	1749	0.720212	1304	45	2.649136	1906	0.777130	1411
20	2.574293	1752	0.721516	1306	50	2.651042	1910	0.778541	1415
25	2.576045	1756	0.722822	1308	55	2.652952	1914	0.779956	1417
30	2.577801		0.724130		132 0	2.654866		0.781373	
		—1760		—1311			—1918		—1421
35	2.579561	1763	0.725441	1313	5	2.656784	1922	0.782794	1423
40	2.581324	1766	0.726754	1316	10	2.658706	1926	0.784217	1425
45	2.583090	1770	0.728070	1318	15	2.660632	1930	0.785642	1429
50	2.584860	1774	0.729388	1320	20	2.662562	1935	0.787071	1431
55	2.586634	1777	0.730708	1323	25	2.664497	1938	0.788502	1434
129 0	2.588411		0.732031		30	2.666435		0.789936	
		—1781		—1326			—1943		—1437
5	2.590192	1784	0.733357	1328	35	2.668378	1947	0.791373	1440
10	2.591976	1788	0.734685	1330	40	2.670325	1951	0.792813	1442
15	2.593764	1792	0.736015	1333	45	2.672276	1955	0.794255	1446
20	2.595556	1795	0.737348	1336	50	2.674231	1960	0.795701	1448
25	2.597351	1799	0.738684	1338	55	2.676191	1964	0.797149	1452
30	2.599150		0.740022		133 0	2.678155		0.798601	
		—1802		—1341			—1968		—1454

TABLE IV. *continued.*

Angle	Log. Mean Mot.	Diff.	Log. Dist.	Diff.	Angle	Log. Mean Mot.	Diff.	Log. Dist.	Diff.
° ′		−1963		−1454	° ′		−2162		−1585
133 5	2.680123		0.800055		136 35	2.766845		0.863874	
10	2.682095	1972	0.801512	1457	40	2.769012	2167	0.865462	1588
15	2.684071	1976	0.802972	1460	45	2.771185	2173	0.867054	1592
20	2.686052	1981	0.804435	1463	50	2.773362	2177	0.868649	1595
25	2.688037	1985	0.805900	1465	55	2.775544	2182	0.870247	1598
30	2.690027	1990	0.807369	1469	137 0	2.777732	2188	0.871849	1602
		−1991		−1472			−2193		−1605
35	2.692021	1998	0.808841	1475	5	2.779925	2198	0.873454	1609
40	2.694019	2003	0.810316	1477	10	2.782123	2203	0.875063	1612
45	2.696022	2007	0.811793	1481	15	2.784326	2208	0.876675	1616
50	2.698029	2011	0.813274	1483	20	2.786534	2214	0.878291	1619
55	2.700040	2016	0.814757	1487	25	2.788748	2218	0.879910	1623
134 0	2.702056	2021	0.816244	1490	30	2.790966	2224	0.881533	1626
5	2.704077	2024	0.817734	1492	35	2.793190	2229	0.883159	1629
10	2.706101	2030	0.819226	1496	40	2.795419	2235	0.884788	1633
15	2.708131	2034	0.820722	1499	45	2.797654	2240	0.886421	1637
20	2.710165	2038	0.822221	1501	50	2.799894	2245	0.888058	1640
25	2.712203	2043	0.823722	1505	55	2.802139	2250	0.889698	1644
30	2.714246	2047	0.825227	1508	138 0	2.804389	2256	0.891342	1647
35	2.716293	2052	0.826735	1511	5	2.806645	2262	0.892989	1651
40	2.718345	2056	0.828246	1514	10	2.808907	2267	0.894640	1654
45	2.720401	2062	0.829760	1517	15	2.811174	2272	0.896294	1659
50	2.722463	2065	0.831277	1520	20	2.813446	2277	0.897953	1661
55	2.724528	2071	0.832797	1524	25	2.815723	2283	0.899614	1666
135 0	2.726599	2075	0.834321	1526	30	2.818006	2289	0.901280	1669
5	2.728674	2080	0.835847	1530	35	2.820295	2294	0.902949	1672
10	2.730754	2084	0.837377	1533	40	2.822589	2300	0.904621	1677
15	2.732838	2089	0.838910	1536	45	2.824889	2306	0.906298	1680
20	2.734927	2094	0.840446	1539	50	2.827195	2311	0.907978	1684
25	2.737021	2099	0.841985	1542	55	2.829506	2316	0.909662	1687
30	2.739120	2104	0.843527	1546	139 0	2.831822	2323	0.911349	1692
35	2.741224	2108	0.845073	1549	5	2.834145	2328	0.913041	1695
40	2.743332	2113	0.846622	1552	10	2.836473	2333	0.914736	1699
45	2.745445	2118	0.848174	1555	15	2.838806	2340	0.916435	1702
50	2.747563	2123	0.849729	1558	20	2.841146	2345	0.918137	1707
55	2.749686	2128	0.851287	1562	25	2.843491	2351	0.919844	1710
136 0	2.751814	2132	0.852849	1565	30	2.845842	2357	0.921554	1714
5	2.753946	2138	0.854414	1569	35	2.848199	2362	0.923268	1718
10	2.756084	2142	0.855983	1571	40	2.850561	2369	0.924986	1722
15	2.758226	2147	0.857554	1575	45	2.852930	2374	0.926708	1726
20	2.760373	2153	0.859129	1578	50	2.855304	2380	0.928434	1729
25	2.762526	2157	0.860707	1582	55	2.857684	2386	0.930163	1734
30	2.764683	2162	0.862289	1585	140 0	2.860070	2392	0.931897	1737

3 S

TABLE IV. *continued.*

Angle.	Log. Mean Mot.	Diff.	Log. Dift.	Diff.	Angle.	Log. Mean Mot.	Diff.	Log. Dift.	Diff.
°		−2392		−1737	° ′		−2667		−1918
140 5	2.862462	2398	0.933634	1741	143 35	2.968661	2675	1.010375	1922
10	2.864860	2405	0.935375	1746	40	2.971336	2681	1.012297	1928
15	2.867265	2410	0.937121	1749	45	2.974017	2689	1.014225	1932
20	2.869675	2416	0.938870	1753	50	2.976706	2696	1.016157	1937
25	2.872091	2422	0.940623	1758	55	2.979402	2703	1.018094	1941
30	2.874513		0.942381		144 0	2.982105		1.020035	
		−2428		−1761			−2711		−1947
35	2.876941	2435	0.944142	1765	5	2.984816	2718	1.021982	1951
40	2.879376	2440	0.945907	1770	10	2.987534	2725	1.023933	1956
45	2.881816	2447	0.947677	1773	15	2.990259	2733	1.025889	1961
50	2.884263	2453	0.949450	1778	20	2.992992	2740	1.027850	1966
55	2.886716	2459	0.951228	1781	25	2.995732	2748	1.029816	1971
141 0	2.889175		0.953009		30	2.998480		1.031787	
		−2466		−1786			−2756		−1976
5	2.891641	2472	0.954795	1790	35	3.001236	2763	1.033763	1980
10	2.894113	2478	0.956585	1794	40	3.003999	2771	1.035743	1986
15	2.896591	2484	0.958379	1799	45	3.006770	2778	1.037729	1991
20	2.899075	2491	0.960178	1802	50	3.009548	2786	1.039720	1996
25	2.901566	2497	0.961980	1807	55	3.012334	2794	1.041716	2000
30	2.904063		0.963787		145 0	3.015128		1.043716	
		−2504		−1811			−2802		−2006
35	2.906567	2510	0.965598	1815	5	3.017930	2809	1.045722	2011
40	2.909077	2517	0.967413	1819	10	3.020739	2818	1.047733	2016
45	2.911594	2523	0.969232	1824	15	3.023557	2825	1.049749	2022
50	2.914117	2530	0.971056	1828	20	3.026382	2833	1.051771	2026
55	2.916647	2536	0.972884	1832	25	3.029215	2841	1.053797	2032
142 0	2.919183		0.974716		30	3.032056		1.055829	
		−2543		−1837			−2850		−2037
5	2.921726	2550	0.976553	1841	35	3.034906	2857	1.057866	2042
10	2.924276	2556	0.978394	1845	40	3.037763	2865	1.059908	2047
15	2.926832	2562	0.980239	1850	45	3.040628	2874	1.061955	2053
20	2.929394	2570	0.982089	1854	50	3.043502	2882	1.064008	2058
25	2.931964	2576	0.983943	1859	55	3.046384	2889	1.066066	2063
30	2.934540		0.985802		146 0	3.049273		1.068129	
		−2584		−1863			−2898		−2069
35	2.937124	2590	0.987665	1867	5	3.052171	2907	1.070198	2074
40	2.939714	2596	0.989532	1872	10	3.055078	2915	1.072272	2080
45	2.942310	2604	0.991404	1877	15	3.057993	2923	1.074352	2085
50	2.944914	2611	0.993281	1881	20	3.060916	2931	1.076437	2090
55	2.947525	2617	0.995162	1885	25	3.063847	2940	1.078527	2096
143 0	2.950142		0.997047		30	3.066787		1.080623	
		−2625		−1890			−2949		−2102
5	2.952767	2631	0.998937	1895	35	3.069736	2957	1.082725	2107
10	2.955398	2639	1.000832	1899	40	3.072693	2965	1.084832	2113
15	2.958037	2645	1.002731	1904	45	3.075658	2974	1.086945	2118
20	2.960682	2653	1.004635	1909	50	3.078632	2983	1.089063	2124
25	2.963335	2659	1.006544	1913	55	3.081615	2992	1.091187	2129
30	2.965994		1.008457		147 0	3.084607		1.093316	
		−2667		−1918			−3000		−2136

TABLE IV. *continued.*

Angle.	Log. Mean Mot.	Diff.	Log. Dift.	Diff.	Angle.	Log. Mean Mot.	Diff.	Log. Dift.	Diff.
147 5	3.087607		1.095452		150 35	3.222156		1.190679	
		3000		2136			3412		2402
10	3.090616	3009	1.097593	2141	40	3.225579	3423	1.193089	2410
15	3.093634	3018	1.099739	2146	45	3.229013	3434	1.195506	2417
20	3.096661	3027	1.101892	2153	50	3.232458	3445	1.197930	2424
25	3.099697	3036	1.104050	2158	55	3.235914	3456	1.200362	2432
30	3.102742	3045	1.106215	2165	151 0	3.239382	3468	1.202801	2439
		3054		2170			3479		2446
35	3.105796	3063	1.108385	2176	5	3.242861	3490	1.205247	2453
40	3.108859	3072	1.110561	2181	10	3.246351	3502	1.207700	2461
45	3.111931	3081	1.112742	2188	15	3.249853	3513	1.210161	2469
50	3.115012	3090	1.114930	2194	20	3.253366	3524	1.212630	2475
55	3.118102	3100	1.117124	2200	25	3.256890	3537	1.215105	2484
148 0	3.121202		1.119324		30	3.260427		1.217589	
		3109		2206			3548		2491
5	3.124311	3118	1.121530	2212	35	3.263975	3559	1.220080	2498
10	3.127429	3128	1.123742	2218	40	3.267534	3572	1.222578	2506
15	3.130557	3137	1.125960	2224	45	3.271106	3583	1.225084	2514
20	3.133694	3146	1.128184	2230	50	3.274689	3596	1.227598	2522
25	3.136840	3156	1.130414	2237	55	3.278285	3607	1.230120	2530
30	3.139996		1.132651		152 0	3.281892		1.232650	
		3166		2243			3620		2537
35	3.143162	3176	1.134894	2249	5	3.285512	3631	1.235187	2545
40	3.146338	3185	1.137143	2255	10	3.289143	3644	1.237732	2553
45	3.149523	3195	1.139398	2262	15	3.292787	3656	1.240285	2562
50	3.152718	3204	1.141660	2268	20	3.296443	3669	1.242847	2569
55	3.155922	3215	1.143928	2274	25	3.300112	3681	1.245416	2577
149 0	3.159137		1.146202		30	3.303793		1.247993	
		3224		2281			3693		2586
5	3.162361	3234	1.148483	2288	35	3.307486	3706	1.250579	2593
10	3.165595	3245	1.150771	2293	40	3.311192	3719	1.253172	2602
15	3.168840	3254	1.153064	2301	45	3.314911	3732	1.255774	2610
20	3.172094	3265	1.155365	2307	50	3.318643	3744	1.258384	2619
25	3.175359	3274	1.157672	2313	55	3.322387	3758	1.261003	2626
30	3.178633		1.159985		153 0	3.326145		1.263629	
		3285		2321			3770		2636
35	3.181918	3296	1.162306	2327	5	3.329915	3783	1.266265	2644
40	3.185214	3305	1.164633	2333	10	3.333698	3797	1.268909	2652
45	3.188519	3316	1.166966	2340	15	3.337495	3810	1.271561	2661
50	3.191835	3327	1.169306	2348	20	3.341305	3823	1.274222	2669
55	3.195162	3337	1.171654	2354	25	3.345128	3836	1.276891	2678
150 0	3.198499		1.174008		30	3.348964		1.279569	
		3347		2360			3850		2687
5	3.201846	3358	1.176368	2368	35	3.352814	3864	1.282256	2696
10	3.205204	3369	1.178736	2375	40	3.356678	3877	1.284952	2705
15	3.208573	3379	1.181111	2381	45	3.360555	3890	1.287657	2713
20	3.211952	3390	1.183492	2389	50	3.364445	3905	1.290370	2722
25	3.215342	3402	1.185881	2396	55	3.368350	3918	1.293092	2732
30	3.218744		1.188277		154 0	3.372268		1.295824	
		3412		2402			3933		2740

TABLE IV. *continued.*

Angle.	Log. Mean Mot.	Diff.	Log. Dist.	Diff.	Angle.	Log. Mean Mot.	Diff.	Log. Dist.	Diff.
° ′		−3933		−2740	° ′		−4611		−3181
154 5	3.376201		1.298564		157 35	3.555279		1.422710	
10	3.380147	3946	1.301314	2750	40	3.559908	4629	1.425904	3194
15	3.384108	3961	1.304073	2759	45	3.564556	4648	1.429110	3206
20	3.388083	3975	1.306841	2768	50	3.569223	4667	1.432328	3218
25	3.392072	3989	1.309619	2778	55	3.573910	4687	1.435559	3231
30	3.396076	4004	1.312405	2786	158 0	3.578615	4705	1.438802	3243
		−4018		−2797			−4725		−3256
35	3.400094	4033	1.315202	2805	5	3.583340	4745	1.442058	3269
40	3.404127	4048	1.318007	2816	10	3.588085	4764	1.445327	3281
45	3.408175	4062	1.320823	2825	15	3.592849	4784	1.448608	3295
50	3.412237	4077	1.323648	2834	20	3.597633	4804	1.451903	3307
55	3.416314	4092	1.326482	2844	25	3.602437	4823	1.455210	3320
155 0	3.420406		1.329326		30	3.607260		1.458530	
		−4108		−2855			−4845		−3334
5	3.424514	4122	1.332181	2864	35	3.612105	4854	1.461864	3347
10	3.428636	4138	1.335045	2874	40	3.616969	4886	1.465211	3360
15	3.432774	4153	1.337919	2883	45	3.621855	4905	1.468571	3374
20	3.436927	4169	1.340802	2894	50	3.626760	4927	1.471945	3388
25	3.441096	4184	1.343696	2905	55	3.631687	4948	1.475333	3401
30	3.445280		1.346601		159 0	3.636635		1.478734	
		−4200		−2914			−4969		−3415
35	3.449480	4216	1.349515	2925	5	3.641604	4991	1.482149	3429
40	3.453696	4232	1.352440	2935	10	3.646595	5012	1.485578	3443
45	3.457928	4247	1.355375	2945	15	3.651607	5033	1.489021	3457
50	3.462175	4264	1.358320	2956	20	3.656640	5056	1.492478	3472
55	3.466439	4280	1.361276	2966	25	3.661696	5078	1.495950	3486
156 0	3.470719		1.364242		30	3.666774		1.499436	
		−4297		−2977			−5099		−3500
5	3.475016	4313	1.367219	2988	35	3.671873	5122	1.502936	3515
10	3.479329	4329	1.370207	2998	40	3.676995	5145	1.506451	3530
15	3.483658	4346	1.373205	3010	45	3.682140	5168	1.509981	3544
20	3.488004	4363	1.376215	3020	50	3.687308	5191	1.513525	3560
25	3.492367	4380	1.379235	3031	55	3.692499	5214	1.517085	3575
30	3.496747		1.382266		160 0	3.697713		1.520660	
		−4397		−3043			−5237		−3589
35	3.501144	4414	1.385309	3053	5	3.702950	5260	1.524249	3606
40	3.505558	4431	1.388362	3065	10	3.708210	5285	1.527855	3621
45	3.509989	4449	1.391427	3076	15	3.713495	5308	1.531476	3636
50	3.514438	4466	1.394503	3087	20	3.718803	5332	1.535112	3652
55	3.518904	4484	1.397590	3099	25	3.724135	5357	1.538764	3668
157 0	3.423388		1.300689		30	3.729492		1.542432	
		−4501		−3111			−5381		−3684
5	3.527889	4519	1.403800	3122	35	3.734873	5406	1.546116	3700
10	3.532408	4538	1.406922	3134	40	3.740279	5431	1.549816	3717
15	3.536946	4555	1.410056	3145	45	3.745710	5456	1.553533	3733
20	3.541501	4574	1.413201	3158	50	3.751166	5481	1.557266	3749
25	3.546075	4593	1.416359	3170	55	3.756647	5507	1.561015	3767
30	3.550668		1.419529		161 0	3.762154		1.564782	
		−4611		−3181			−5533		−3783

TABLE IV. *continued.*

Angle.	Log. Mean Mot.	Diff.	Log. Diff.	Diff.
		−5533		−3783
161 5	3.767687		1.568565	
		5559		3800
10	3.773246		1.572365	
		5585		3817
15	3.778831		1.576182	
		5611		3835
20	3.784442		1.580017	
		5628		3852
25	3.790080		1.583869	
		5665		3869
30	3.795745		1.587738	
		−5692		−3888
35	3.801437		1.591626	
		5720		3905
40	3.807157		1.595531	
		5747		3923
45	3.812904		1.599454	
		5775		3942
50	3.818679		1.603396	
		5804		3960
55	3.824483		1.607356	
		5832		3979
162 0	3.830315		1.611335	
		−5860		−3998
5	3.836175		1.615333	
		5890		4016
10	3.842065		1.619349	
		5918		4036
15	3.847983		1.623385	
		5948		4055
20	3.853931		1.627440	
		5978		4074
25	3.859909		1.631514	
		6008		4094
30	3.865917		1.635608	
		−6038		−4114
35	3.871955		1.639722	
		6069		4134
40	3.878024		1.643856	
		6100		4154
45	3.884124		1.648010	
		6131		4175
50	3.890255		1.652185	
		6162		4195
55	3.896417		1.656380	
		6194		4216
163 0	3.902611		1.660596	
		−6226		−4237
5	3.908837		1.664833	
		6258		4258
10	3.915095		1.669091	
		6291		4280
15	3.921386		1.673371	
		6324		4301
20	3.927710		1.677672	
		6358		4324
25	3.934068		1.681996	
		6392		4345
30	3.940460		1.686341	
		−6425		−4368
35	3.946885		1.690709	
		6459		4390
40	3.953344		1.695099	
		6494		4413
45	3.959838		1.699512	
		6530		4436
50	3.966368		1.703948	
		6565		4459
55	3.972933		1.708407	
		6600		4482
164 0	3.979533		1.712889	
		−6637		−4507
5	3.986170		1.717396	
		6673		4530
10	3.992843		1.721902	
		6710		4555
15	3.999553		1.726481	
		6747		4579
20	4.006300		1.731060	
		6785		4603
25	4.013085		1.735663	
		6823		4629
30	4.019908		1.740292	
		−6861		−4654

Angle.	Log. Mean Mot.	Diff.	Log. Diff.	Diff.
		−6861		−4654
164 35	4.026769		1.744946	
		6900		4680
40	4.033669		1.749626	
		6940		4705
45	4.040609		1.754331	
		6979		4731
50	4.047588		1.759062	
		7019		4758
55	4.054607		1.763820	
		7060		4785
165 0	4.061667		1.768605	
		−7101		−4811
5	4.068768		1.773416	
		7142		4839
10	4.075910		1.778255	
		7185		4866
15	4.083095		1.783121	
		7227		4894
20	4.090322		1.788015	
		7269		4923
25	4.097591		1.792938	
		7313		4950
30	4.104904		1.797888	
		−7357		−4980
35	4.112261		1.802868	
		7401		5009
40	4.119662		1.807877	
		7446		5038
45	4.127108		1.812915	
		7491		5069
50	4.134599		1.817984	
		7537		5098
55	4.142136		1.823082	
		7584		5129
166 0	4.149720		1.828211	
		−7630		−5160
5	4.157350		1.833371	
		7678		5191
10	4.165028		1.838562	
		7727		5223
15	4.172755		1.843785	
		7775		5255
20	4.180530		1.849040	
		7824		5288
25	4.188354		1.854328	
		7874		5320
30	4.196228		1.859648	
		−7924		−5353
35	4.204152		1.865001	
		7976		5388
40	4.212128		1.870389	
		8027		5421
45	4.220155		1.875810	
		8080		5456
50	4.228235		1.881266	
		8132		5490
55	4.236367		1.886756	
		8185		5526
167 0	4.244553		1.892282	
		−8241		−5562
5	4.252794		1.897844	
		8296		5599
10	4.261090		1.903443	
		8352		5635
15	4.269442		1.909078	
		8408		5672
20	4.277850		1.914750	
		8465		5710
25	4.286315		1.920460	
		8520		5748
30	4.294838		1.926208	
		−8582		−5788
35	4.303420		1.931996	
		8642		5826
40	4.312062		1.937822	
		8701		5866
45	4.320763		1.943688	
		8763		5907
50	4.329526		1.949595	
		8825		5947
55	4.338351		1.955542	
		8888		5989
168 0	4.347239		1.961531	
		−8951		−6031

TABLE IV. *continued*.

Angle.	Log. Mean Mot.	Diff.	Log. Dist.	Diff.	Angle.	Log. Mean Mot.	Diff.	Log. Dist.	Diff.
		8951		6031			12745		8542
168° 5	4.356190	9016	1.967562	6073	171° 35	4.804629	12873	2.268806	8627
10	4.365206	9082	1.973635	6117	40	4.817502	13004	2.277433	8715
15	4.374288	9147	1.979752	6161	45	4.830506	13139	2.286148	8803
20	4.383435	9215	1.985913	6205	50	4.843645	13274	2.294951	8894
25	4.392650	9284	1.992118	6250	55	4.856919	13414	2.303845	8986
30	4.401934	9352	1.998368	6296	172 0	4.870333	13557	2.312831	9081
35	4.411286	9423	2.004664	6342	5	4.883890	13701	2.321912	9177
40	1.420709	9494	2.011006	6390	10	4.897591	13850	2.331089	9275
45	4.430203	9567	2.017396	6437	15	4.911441	14002	2.340364	9376
50	4.439770	9640	2.023833	6486	20	4.925443	14156	2.349740	9479
55	4.449410	9714	2.030319	6535	25	4.939599	14314	2.359219	9584
169 0	4.459124	9790	2.036854	6585	30	4.953913	14476	2.368803	9691
5	4.468914	9857	2.043439	6636	35	4.968389	14642	2.378494	9801
10	4.478781	9945	2.050075	6688	40	4.983031	14811	2.388295	9914
15	4.488726	10024	2.056763	6740	45	4.997842	14984	2.398209	10028
20	4.498750	10104	2.063503	6792	50	5.012826	15160	2.408237	10146
25	4.508854	10186	2.070295	6847	55	5.027986	15342	2.418383	10266
30	4.519040	10269	2.077142	6902	173 0	5.043328	15528	2.428649	10390
35	4.529309	10353	2.084044	6958	5	5.058856	15718	2.439039	10516
40	4.539662	10438	2.091002	7014	10	5.074574	15842	2.449555	10645
45	4.550100	10526	2.098016	7072	15	5.090486	16112	2.460200	10778
50	4.560626	10614	2.105088	7130	20	5.106598	16316	2.470978	10913
55	4.571240	10704	2.112218	7190	25	5.122914	16526	2.481891	11053
170 0	4.581944	10796	2.119408	7250	30	5.139440	16741	2.492944	11196
5	4.592740	10888	2.126658	7312	35	5.156181	16962	2.504140	11343
10	4.603628	10982	2.133970	7374	40	5.173143	17188	2.515483	11493
15	4.614610	11079	2.141344	7438	45	5.190331	17421	2.526976	11648
20	4.625689	11176	2.148782	7503	50	5.207752	17660	2.538624	11806
25	4.636865	11276	2.156285	7568	55	5.225412	17905	2.550430	11970
30	4.648141	11377	2.163853	7636	174 0	5.243317	18157	2.562400	12137
35	4.659518	11480	2.171489	7703	5	5.261474	18416	2.574537	12310
40	4.670998	11585	2.179192	7773	10	5.279890	18684	2.586847	12487
45	4.682583	11691	2.186965	7844	15	5.298574	18958	2.599334	12670
50	4.694274	11799	2.194809	7916	20	5.317532	19240	2.612004	12858
55	4.706073	11910	2.202725	7988	25	5.336772	19532	2.624862	13051
171 0	4.717983	12023	2.210713	8064	30	5.356304	19832	2.637913	13252
5	4.730006	12138	2.218777	8139	35	5.376136	20142	2.651165	13456
10	4.742144	12254	2.226916	8217	40	5.396278	20460	2.664621	13669
15	4.754398	12373	2.235133	8296	45	5.416738	20789	2.678290	13889
20	4.766771	12495	2.243429	8377	50	5.437527	21130	2.692179	14114
25	4.779266	12618	2.251806	8458	55	5.458657	21480	2.706293	14348
30	4.791884	12745	2.260264	8542	175 0	5.480137	21843	2.720641	14589

TABLE IV. *continued.*

Angle.	Log. Mean Mot.	Diff.	Log. Diff.	Diff.	Angle.	Log. Mean Mot.	Diff.	Log. Diff.	Diff.
° ′					° ′				
		21843		14·89			44143		29442
175 5	5.501980		2.735230		177 35	6.426134		3.351936	
		22219		14839			45694		30476
10	5.524199		2.750069		40	6.471828		3.382412	
		22607		15098			47357		31584
15	5.546806		2.765167		45	6.519185		3.413996	
		23009		15365			49147		32777
20	5.569815		2.780532		50	6.568332		3.446773	
		23425		15642			51077		34063
25	5.593240		2.796174		55	6.619409		3.480836	
		23857		15930			53164		35453
30	5.617097		2.812104		178 0	6.672573		3.516289	
		24304		16227			55428		36964
35	5.641401		2.828331		5	6.728001		3.553253	
		24770		16537			57895		38607
40	5.666171		2.844868		10	6.785896		3.591860	
		25253		16859			60590		40403
45	5.691424		2.861727		15	6.846486		3.632263	
		25754		17192			63549		42375
50	5.717178		2.878919		20	6.910035		3.674638	
		26276		17540			66812		44550
55	5.743454		2.896459		25	6.976847		3.719188	
		26820		17903			70426		46960
176 0	5.770274		2.914362		30	7.047273		3.766148	
		27387		18279			74454		49644
5	5.797661		2.932641		35	7.121727		3.815792	
		27978		18673			78972		52655
10	5.825639		2.951314		40	7.200699		3.868447	
		28594		19084			84072		56055
15	5.854233		2.970398		45	7.284771		3.924502	
		29239		19513			89876		59925
20	5.883472		2.989911		50	7.374647		3.984427	
		29912		19961			96542		64367
25	5.913384		3.009872		55	7.471189		4.048794	
		30619		20432			104275		69522
30	5.944003		3.030304		179 0	7.575464		4.118316	
		31358		20925			113355		75576
35	5.975361		3.051229		5	7.688819		4.193892	
		32135		21441			124168		82783
40	6.007496		3.072670		10	7.812987		4.276675	
		32950		21985			137264		91511
45	6.040446		3.094655		15	7.950251		4.368189	
		33808		22556			153452		102305
50	6.074254		3.117211		20	8.103703		4.470494	
		34711		23158			173967		115981
55	6.108965		3.140369		25	8.277670		4.586475	
		35664		23793			200834		133893
177 0	6.144629		3.164162		30	8.478504		4.720368	
		36670		24463			237539		158362
5	6.181299		3.188625		35	8.716043		4.878730	
		37736		25173			290726		193819
10	6.219035		3.213798		40	9.006769		5.072549	
		38865		25925			374813		249877
15	6.257900		3.239723		45	9.381582		5.322426	
		40062		26723			528271		352182
20	6.297962		3.266446		50	9.909853		5.674608	
		41336		27572			903089		602060
25	6.339298		3.294018		55	10.812942		6.276668	
		42693		28476	180 0					
30	6.381991		3.322494							
		44143		29442						

TABLE V. *Of Abscissæ and Ordinates.*

Abscissæ.	Ordinates.	Abscissæ.	Ordinates.	Abscissæ.	Ordinates.
0,125	0,70710	18	8,4853	48	13,8564
0,25	1,00000	19	8,7178	49	14,0000
0,50	1,41421	20	8,9444	50	14,1422
0,75	1,73205	21	9,1651	52	14,4222
1,00	2,00000	22	9,3808	56	14,9666
1,5	2,44950	23	9,5916	60	15,4920
2,0	2,82843	24	9,7979	64	16,0000
2,5	3,16225	25	10,0000	68	16,4924
3,0	3,46410	26	10,1980	72	16,9706
3,5	3,74165	27	10,3923	76	17,4356
4,0	4,00000	28	10,5830	80	17,8888
4,5	4,24265	29	10,7703	81	18,0000
5,0	4,47210	30	10,9545	84	18,3302
5,5	4,69040	31	11,1355	88	18,7616
6,0	4,89900	32	11,3138	92	19,1832
6,5	5,09900	33	11,4891	96	19,5958
7,0	5,29150	34	11,6619	100	20,0000
7,5	5,47725	35	11,8322	104	20,3960
8,0	5,65690	36	12,0000	108	20,7846
8,5	5,83095	37	12,1655	112	21,1660
9,0	6,00000	38	12,3288	116	21,5406
9,5	6,16400	39	12,4900	120	21,9090
10,0	6,3245	40	12,6490	121	22,0000
11	6,6332	41	12,8062	124	22,2710
12	6,9282	42	12,9615	128	22,6276
13	7,2111	43	13,1149	132	22,9782
14	7,4833	44	13,2664	136	23,3238
15	7,7460	45	13,4164	140	23,6644
16	8,0000	46	13,5647	144	24,0000
17	8,2462	47	13,7113		

F I N I S.

A P P E N D I X.

CHAPT. 6, § 4, *Page* 30; *add as follows.*

LAMBERT in his work intituled " *Orbitarum Cometicarum Proprietates infigniores,*" gives a more accurate and much more expeditious method of drawing a Parabola, founded on the following Theorem.

T H E O R E M.

The point of concourfe of any Tangent to a Parabola, and a line drawn from the Focus at right Angles to the Tangent, will ever be in the Vertical Tangent. Or let V C M, (Fig. 20.) be a Parabola, whofe Vertex is V; its Focus S; its Axis E V A, and Vertical Tangent F B V. Draw at pleafure the Tangent G C B E, meeting the Curve at C, and the Axis produced at E; and from the Focus the Line S B, at Right Angles to the Tangent G C B E. Then the point of concourfe of thefe two Lines will ever be in the Line F B V.

For from the point of contact C, draw the Ordinate to the Axis C D; and the Radius Vector C S. Then by the property of the Parabola, we have E V $=$ V D. Therefore by fimilar Triangles E B $=$ B C. Therefore C E, is biffected by F B V, in the point B. Again, by the Parabola we have S E $=$ S C. And S B, at Right Angles to C E, by Conftruction. There-

a

fore C E, is biffected alfo by S B, in the point B. Therefore the point of concourfe of the Tangent C E, and Line S B, is in the Vertical Tangent F B V. Q.E.D.

Hence if the Axis and Focal Diftance of a Parabola be given; the Curve is eafily defcribed by its Tangents. Thus. From the Vertex V, (Fig. 20.) draw the Vertical Tangent V F′, at Right Angles to the Axis; then let one fide of a Square pafs through the Focus S, and its right Angle touch the Line V F′, in any points B, B′, B″, &c. and by the other fide of the Square, draw B G, B′ G′, B″ G″, &c.; then thefe lines will be Tangents to the Curve: which as the Points B, B′, may be taken near each other at pleafure, will circumfcribe the Parabola to any required accuracy; and the Curve thus drawn on Pafteboard, may be eafily cut out and transferred to any drawing.

But whatever method be ufed for the defcription of the Curve, it will be very proper to try it in different places by the Abfciffæ and Ordinates. This will be a fecurity againft any errors which may creep into the drawing, and will by the help of Table 5, be attended with very little trouble. In this Table the Focal Diftance is taken as 1; and the lengths of the Ordinates for any Parabola, may be eafily found on the Sector, by making the Focal Diftance a tranfverfe to 1, 1, on the Line of Lines. The Ordinates thus found muft be fet off on perpendiculars to the Axis, drawn from their correfponding Abfciffæ; beginning at the Vertex. A fmall number of points thus found, will keep the drawing right. Indeed (as Mr. Barker obferves) in thofe Parabolas whofe Focal Diftance is very fmall; the parts of the Curve diftant from the Vertex are fo near right Lines, and the Angle from the Focus varies fo flowly, that perhaps the eafieft and moft accurate method in thefe cafes will be to draw the whole Curve by its Abfciffæ and Ordinates; and for that reafon the table is carried much farther than is generally wanted.

CHAPT. 8, § 6, *Page* 43; *add as follows.*

IT fometimes though very rarely happens, that the Orbit of a Comet is fo placed with refpect to that of the Earth, that its motion during the whole time of its appearance is in a great Circle paffing through or very near, the Poles of the Ecliptic. Its motion in longitude is therefore either nothing, or fo very fmall, that it is impoffible to deduce from it the curtate diftances from the Earth, by the proportions, $T P = \dfrac{t'' \operatorname{Sin.} m'}{t' \operatorname{Sin.} m''} \times T' P'$

and $T'' P'' = \dfrac{t'' \operatorname{Sin.} m}{t \operatorname{Sin.} m''} \times T' P'$.

This uncommon cafe occurred in the Comet which was obferved in the month of January of this Year, (1793). In order therefore in fuch circumftances, to obtain the Diftances of the Comet from the Earth, m, m', and m'', muft reprefent the motions of the Comet in Latitude; and the fuppofition of the Diftance of the Comet from the Earth at the middle Obfervation, muft be not of the Curtate Diftance $T' P'$, but of the real Diftance $T' C'$. Hence by the proportions $T C = \dfrac{t'' \operatorname{Sin.} m'}{t' \operatorname{Sin.} m''} \times T' C'$,

and $T'' C'' = \dfrac{t'' \operatorname{Sin.} m}{t \operatorname{Sin.} m''} \times T' C'$, $T C$, and $T'' C''$; will be found.

4

Thefe being fet off on the Lines T F, and T F″, (Fig. 5.) as to C, and C″; perpendiculars let fall on the Line T B, will give on that Line, the Lengths of T P, and T″ P″; which being transferred to the Lines T E, and T″ E″; of Fig. 4, the Lengths of S P, and S P″; S C, and S C″; P P″, and C C″; are found precifely as in the former method. In this cafe no correction is required for the places either of the Earth or Comet, at the middle Obfervation.

ERRATA and CORRECTIONS.

Table of Contents, Page ii. Line 2, for Declination, read Delineation.

Page 8, Line 11, for c G, read c C.

Page 12, Line 4, for E T = E D, read E T — E D.

Page 14, Line 2, for S D² + S F², read S D² × S F².

Page 21, for § 5, read § 6, and fo on to the end of that Chapter.

Page 24, Line 6, for by Tangents drawn, &c. read by Lines drawn from T″, and C″, parallel with the Radii S T, and S C, and cut by Tangents drawn from T and C.

Page 45, Line 9, for S′ C′, read S P′.

Page 47, for § 17, read § 13, and fo on to the end of that Chapter.

Page 58, Line 6, for V u V, read V u V′.

Page 72, Line 13, for or by the other Table of the Parabolic fall of Comets, read or by Barker's Table of the Parabola.

Page 137, Line 2, for Formula, read Formulæ.

Page 152, Line laſt, for Second, read Firſt.

Page 156, Line 2, for δ³ 6, read δ³ 6′.

Page 157, Line 13, for i i, read i i′.

Page 168, Lines 16, 20, and 21, for 0,85861, read 0,858651.

Page 182, Line 9, for Second, read Firſt.

Page 186, Line 17, for m — m, read m — m′.

Page 203, Line 1, for Table 1, read Table 3.

Table 3, oppoſite Days 7100, for Anomaly 161°.44′.41,″1, read 161°, 43′.41″, 1.

Table 4, oppoſite Angle 18°.5′, in the Column of Log. Diſt. for 0.010800, read 0.010860.

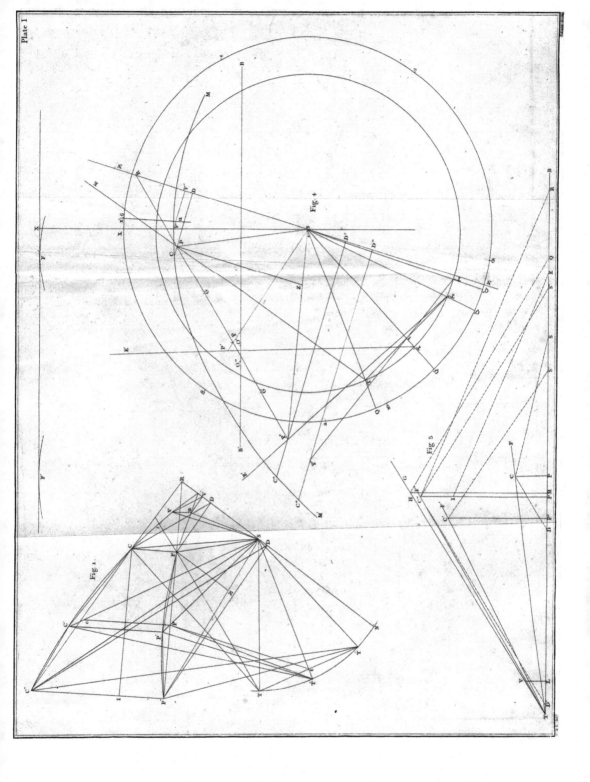

The material originally positioned here is too large for reproduction in this reissue. A PDF can be downloaded from the web address given on page iv of this book, by clicking on 'Resources Available'.

Plate II

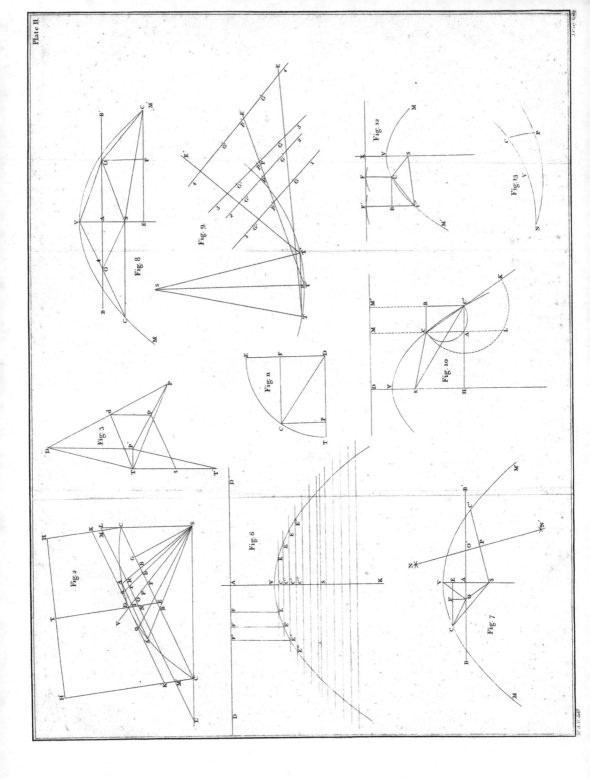

The material originally positioned here is too large for reproduction in this reissue. A PDF can be downloaded from the web address given on page iv of this book, by clicking on 'Resources Available'.

Plate III

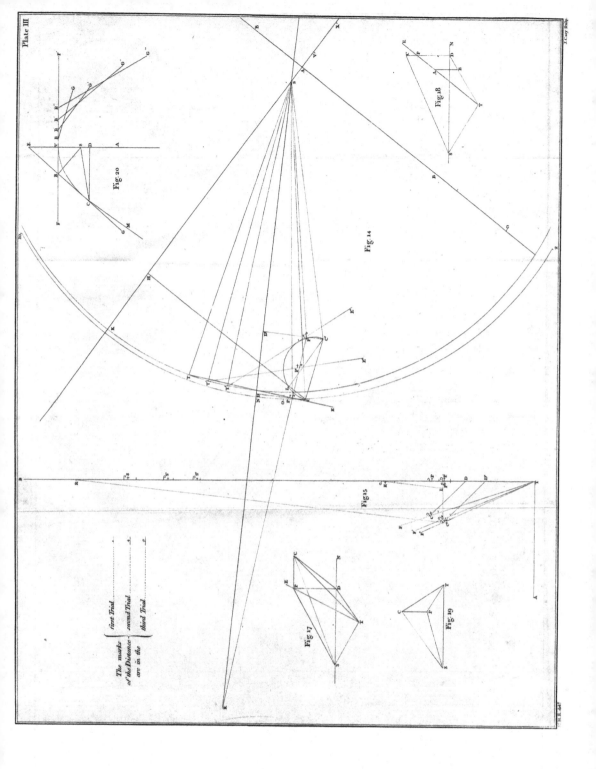

The material originally positioned here is too large for reproduction in this reissue. A PDF can be downloaded from the web address given on page iv of this book, by clicking on 'Resources Available'.

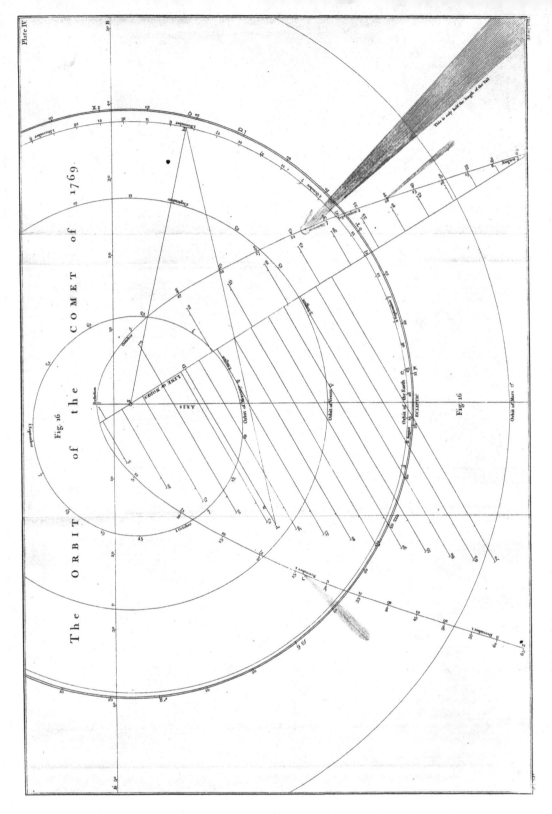

The material originally positioned here is too large for reproduction in this reissue. A PDF can be downloaded from the web address given on page iv of this book, by clicking on 'Resources Available'.